Palgrave Studies in Impact Finance

Series Editor
Mario La Torre, Department of Management, Sapienza University of
Rome, Rome, Italy

The *Palgrave Studies in Impact Finance* series provides a valuable scientific 'hub' for researchers, professionals and policy makers involved in Impact finance and related topics. It includes studies in the social, political, environmental and ethical impact of finance, exploring all aspects of impact finance and socially responsible investment, including policy issues, financial instruments, markets and clients, standards, regulations and financial management, with a particular focus on impact investments and microfinance.

Titles feature the most recent empirical analysis with a theoretical approach, including up to date and innovative studies that cover issues which impact finance and society globally.

Chiara Mio · Marisa Agostini ·
Francesco Scarpa

Sustainability Reporting

Conception, International Approaches and Double
Materiality in Action

Chiara Mio
Department of Management—Venice
School of Management, Sustainability
Lab
Ca' Foscari University
Venice, Italy

Marisa Agostini
Department of Management—Venice
School of Management, Sustainability
Lab
Ca' Foscari University
Venice, Italy

Francesco Scarpa
Department of Management—Venice
School of Management, Sustainability
Lab
Ca' Foscari University
Venice, Italy

ISSN 2662-5105   ISSN 2662-5113   (electronic)
Palgrave Studies in Impact Finance
ISBN 978-3-031-58451-0   ISBN 978-3-031-58449-7   (eBook)
https://doi.org/10.1007/978-3-031-58449-7

Cover illustration: © John Rawsterne/patternhead.com

This Palgrave Macmillan imprint is published by the registered company Springer Nature Switzerland AG
The registered company address is: Gewerbestrasse 11, 6330 Cham, Switzerland

If disposing of this product, please recycle the paper.

CONTENTS

ABBREVIATIONS

CSR	Corporate Social Responsibility
CSRD	Corporate Sustainability Reporting Directive
DA	Dialogic Accounting
EC	European Commission
EFRAG	European Financial Reporting Advisory Group
ESG	Environmental, Social and Governance
ESRS	European Sustainability Reporting Standards
NFI	Non-Financial Information
NFRD	Non-Financial Reporting Directive
SME (SMEs)	Small and Medium-sized Enterprise (Small and Medium-sized Enterprises)

LIST OF TABLES

CHAPTER 1

Introduction

Abstract This chapter provides a general introduction to sustainability reporting and non-financial disclosure and outlines the book's objectives, benefits, and target audiences. It initiates with a discussion on the role and significance of sustainability reporting within the broader landscape of corporate reporting. The goals of the book are then articulated, emphasizing its dual focus on regulatory aspects and the academic discourse, as well as its particular emphasis on materiality. Subsequently, the section discusses how the book can be a valuable resource for scholars, researchers, students, companies, and practitioners to achieve a deeper understanding of the intricate dynamics within sustainability reporting and its ever-evolving regulatory landscape. The section concludes by briefly outlining the book's structure and the content of each chapter.

Keywords Sustainability reporting · Non-financial disclosure ·
Sustainability accounting · Corporate reporting

Sustainability accounting and reporting have emerged as central topics in both corporate practice and academic discourse. In light of urgent social and environmental issues, such as climate change and social inequality, companies are increasingly expected to take responsibility for their social and environmental impacts and to consider how these issues can significantly affect their business. Sustainability accounting and reporting act as essential tools for companies to manage and monitor these impacts, providing stakeholders with assurance that sustainability concerns are being effectively addressed. Moreover, the landscape of sustainability reporting is evolving rapidly, with a rising number of companies participating, an expansion of topics to be disclosed, and the introduction of new standards and regulations.

The motivation and the rationale of this book move from three main premises.

The first premise is that, although traditional financial accounting is a valuable tool for companies and stakeholders, it fails to provide a comprehensive and reliable overview of a company's potential for growth and development, the process of value creation, and its overall performance. To be relevant, accounting should provide users with the information they need to achieve an in-depth understanding of firm value, where "value" should be conceived as both effective and potential, as both created and destroyed, and both from a financial and non-financial point of view (social, environmental, relational, intellectual, etc.). For this reason, academics and practitioners have pointed out the shortcomings of financial accounting information and indicators, because they tend to focus on the ability of the firm to maximize shareholder value in the short term. In other terms, financial accounting proves to be, at least partially, inadequate if confronted with more recent understandings of the purpose of the firm, which is to create value for all stakeholders (and not only shareholders) in order to support sustainable development. Therefore, stakeholders increasingly demand information to understand and assess the process of (stakeholder) value creation in order to decide whether to establish (or continue) business relationship with a firm. Faced with the pressure to be accountable for how they create value for all stakeholders, firms are expected to disclose information covering the different perspectives from which the process of stakeholder value creation can be observed and measured, which correspond to the three pillars of sustainability: economic, environmental, and social. In the last decades, sustainability reporting, i.e., the practice to disclose sustainability-related impacts, risks, and opportunities to uncover the process of stakeholder value creation, has become an established corporate reporting practice, although with different degrees of importance depending on the size and the institutional context of companies. The KMPG Survey of Sustainability Reporting (KPMG, 2022) shows that the 96% of the world's 250 largest companies by revenue report on sustainability or ESG matters in 2022, while the rate of reporting was 35% at the beginning of 2000s.

The second premise of this book is that the regulation of sustainability reporting is undergoing profound transformations that are radically changing the nature of this practice and how it is approached by companies. In contrast to the well-established and regulated realm of financial accounting and reporting, sustainability reporting has traditionally been characterized by a more loosely structured context, where companies had the autonomy to decide, at their discretion, the nature and extent of non-financial information to disclose. Driven by stakeholder demands to be provided with more reliable, comparable, and relevant non-financial information, the practice of sustainability reporting has been increasingly regulated. This regulation has taken two main forms. On the one hand, several voluntary frameworks and standards of sustainability reporting have been developed by dedicated standard-setter bodies, such as the Global Reporting Initiative (GRI), the Sustainability Accounting Standards Board (SASB), and, more recently, the International Sustainability Standards Board (ISSB) under the auspices of the IFRS Foundation. The lack of convergence among these frameworks contributes to aggravate the complexity of the field. On the other hand, the European Union has been the precursor in mandatory regulation in the field of sustainability reporting, requiring certain companies to prepare and publish a sustainability report. Mandatory sustainability reporting was introduced by the EU Non-Financial Reporting Directive (NFRD, 2014) that has been recently reviewed by the EU Corporate Sustainability Reporting Directive (CSRD, 2022).

The third premise is that the ever-increasing importance of sustainability reporting and the deep transformations taking place in the regulatory landscape carry significant implications for all the actors involved in this domain. Principally, businesses and practitioners face the challenge to navigate an environment marked by continual evolution. The escalating demand for non-financial information has made sustainability reporting a pivotal determinant in shaping the quality of firm relationships with stakeholders (e.g., customers, suppliers, investors, lenders, governments, NGOs, etc.) which, in turn, can influence the overall success and longevity of a business. Furthermore, many businesses find themselves in the position of having to engage in sustainability reporting for the first time or to introduce significant changes in light of the new regulatory frameworks. Additionally, academics and scholars are in the midst of a complex and ever-changing subject. The field of sustainability reporting not only presents challenges but also offers a plethora of opportunities for research. Navigating through this dynamic landscape requires a keen understanding of the latest developments and emerging trends, making it a stimulating and demanding area of academic exploration. Lastly, accounting students, as future professionals in this field, need reliable and well-structured guidance to approach and comprehend the multifaceted changes occurring in the realm of sustainability reporting.

Based on these premises, the overall purpose of this book is to provide a comprehensive overview of the evolution, the state-of-the-art, and the future directions of sustainability reporting. To achieve this overall purpose, the book is guided by three specific and interconnected aims. First, the book aims to investigate the evolution of sustainability reporting regulation (i.e., the shift from a voluntary to a mandatory approach) and to explore the variety of frameworks and standards, normative sources and regulatory initiatives aimed at promoting and standardizing sustainability reporting at the international level. Second, the book aims to provide a systematic review of academic literature on sustainability reporting and non-financial disclosure to consolidate existing knowledge in the field and suggest directions for future developments. Third, the book examines the concept of materiality in sustainability reporting and provides an empirical analysis of the quantity and quality of materiality disclosures in sustainability reporting across the globe.

Overall, the book aspires to be a comprehensive and reader-friendly guide, offering insights into the past, present, and future of sustainability reporting and its regulatory landscape. The book aims to bridge the gap between theoretical insights and practical implications by approaching the topic of sustainability reporting from academic and professional perspectives. This dual perspective enhances the book's accessibility, making it equally beneficial for academic scholars exploring the intricacies of sustainability reporting and companies and practitioners navigating the regulatory changes.

To achieve its purposes the book is organized into four main chapters.

Chapter 2 provides an introduction to sustainability reporting, by presenting the process of reporting, building on the hierarchical staged process model (Deegan & Unerman, 2011). According to this model, the sustainability reporting process includes five stages, which involve decisions related to the following issues: (1) Why does the entity disclose sustainability information? (2) Who are the stakeholders to whom the sustainability report is directed? (3) What information and issues should be included in the sustainability report? (4) How should the sustainability report be prepared? (5) Should the sustainability report be assured?

Chapter 3 explores the historical evolution and current landscape of sustainability reporting regulation. The first part deals with the origins, drivers, and initial development of non-financial disclosure and sustainability reporting in order to explain why these practices have become increasingly important over the years. The chapter discusses how the concept of corporate value creation has been redefined to find a balance between economic growth and non-financial concerns, taking into account the interests of all stakeholders (not only shareholders). The stakeholder-oriented approach and the process of stakeholder value creation have gained prominence in the development of corporate strategies, influencing the type of information companies are expected to disclose. Subsequently, the chapter explores the evolution and the state-of-the-art in the regulation of sustainability reporting across the globe. In turn, realizing the importance of the issue, many regulatory bodies have put companies under pressure to operate according to sustainable principles and criteria and to adequately monitor this behavior. A special focus in the chapter is paid to the European Union's transition from a voluntary to a mandatory approach to sustainability reporting. The chapter presents the actions through which the European Commission (EC) initially encouraged companies to disclose non-financial information on a voluntary basis (using recommendations and communications without any binding effect), before moving to legislation (i.e., the NFRD and CSRD) marking the transition to mandatory sustainability reporting. Furthermore, the chapter explores the peculiarities of the market-driven regulatory context in the USA and investigates other significant international settings (UK, Australia, New Zealand India, Asia, and South Africa). This analysis offers insights into the diversity of regulatory frameworks shaping non-financial disclosure and sustainability reporting across the globe.

Chapter 4 offers a systematic review of the literature on non-financial disclosure and sustainability reporting. This chapter contributes to broaden our understanding of non-financial disclosure and sustainability reporting by examining its historical evolution, categorizing existing knowledge into different research streams, providing a descriptive analysis of the literature in this field, and bringing to the forefront the most recent empirical findings regarding the drivers and consequences of sustainability reporting. The review shows how literature has emphasized the importance of NFI's quality and transparency to boost stakeholder trust (Aureli et al., 2019; La Torre et al., 2018) and that a particular focus in the academic debate is placed on the role of regulation in compelling businesses to enhance transparency on sustainability matters. Although some research indicates that regulation is not always an effective means to change corporate behaviors and practices (Bebbington et al., 2012; Luque-Vilchez & Larrinaga, 2016), some attempts have been made (especially across European Union) through the issuance of mandatory requirements for sustainability reporting. Therefore, literature has discussed the coexistence of mandatory and voluntary disclosure and the associated issues (Bold, 2017; Delbard, 2008; Doshi et al., 2013; Martin-Sardesai & Guthrie, 2019), highlighting the ambiguous status of sustainability reporting, which is now considered mandatory in nature (CSRD, 2022), but still voluntary in terms of being driven by stakeholders' expectations (O'Donovan, 2002) and firm characteristics (Clarkson et al., 2008). By systematically reconstructing the academic debate, the book not only captures the current state of knowledge on sustainability reporting but also serves as a guide for developments, identifying largely unexplored research areas and questions that are still seeking answers.

Chapter 5 focuses on the principle of materiality, recognizing its significance in shaping the nature and content of sustainability reports. First, this chapter explores the roots of materiality in financial accounting, before moving to examine how this concept has evolved and adapted to the specific needs of sustainability reporting. Subsequently, the chapter investigates the variety of materiality approaches (i.e., impact materiality, financial materiality, double materiality, and dynamic materiality) adopted in the most important sustainability reporting frameworks and standards. Finally, the chapter concludes with an empirical analysis that investigates the level of corporate disclosure regarding the principle of materiality and the materiality assessment process. Utilizing a sample of companies within the EURO STOXX 50 and the S&P 50, we assessed the extent to which companies communicate the use of the materiality principle and provide details on the materiality assessment process in their sustainability reporting. Through this approach, the chapter aims to provide a comprehensive overview of materiality, intertwining its historical evolution with its current manifestations in the context of sustainability reporting. The special focus that this book dedicates to the issue of materiality is motivated by its role as an overarching principle in sustainability reporting. The selected approach to materiality is closely linked to the objective and function that are attributed to sustainability reporting, which can range from an ethical interpretation (impact materiality) to a more strategic vision (financial materiality). Furthermore, the approach to materiality affects the target audience of the report, which can be identified solely as investors (financial materiality) or include all categories of stakeholders (impact and double materiality). How materiality is interpreted, then, influences the entire architecture of the reporting framework and impacts the information and KPIs it requires to report. Being such an overarching sustainability reporting principle, sustainability materiality can be seen as a lens through which the convergence/divergence between the variety of regulatory and normative sustainability reporting frameworks and standards can be examined and the key differences and commonalities can be identified.

The **final section** offers concluding remarks that summarize key insights and contributions to research and practice. This section discusses the primary benefits of the book, its contributions to understanding sustainability reporting, and the implications for research and practice. It also acknowledges the limitations of our research and provides suggestions for future research directions.

The book caters to diverse audiences, spanning both academic and business contexts. In the academic realm, the book stands as a valuable resource for scholars, researchers, and students, fostering a deeper understanding of the intricate dynamics within sustainability reporting and its ever-evolving regulatory landscape, while also opening avenues for future developments.

In the business context, the book addresses the needs of companies at various stages of engagement with sustainability. For those just entering the sustainability arena, it serves as a foundational guide, elucidating the motivations and benefits of high-quality sustainability reporting and offering practical insights into the international best practices to prepare a sustainability report. For companies already well-versed in sustainability, the book acts as a strategic tool, keeping them informed about the ongoing regulatory changes and positioning them to navigate challenges and to capitalize on existing opportunities in the dynamic realm of sustainability reporting. In essence, the book caters to a diverse corporate audience, making it an indispensable resource for those taking their initial steps into sustainability reporting and a reliable tool for companies aiming to stay at the forefront of developments. Its readability and comprehensive approach make it a valuable asset, bridging the knowledge gap and empowering companies to navigate the complex landscape of sustainability with confidence and foresight.

By bridging the academic-business divide, the book fosters a holistic understanding of sustainability reporting. Its dual focus ensures that it contributes both to the academic discourse and to the practical knowledge required by businesses aiming to integrate sustainability effectively into their operations and reporting practices.

This book plays a pivotal role in nurturing a comprehensive understanding of sustainability reporting, connecting the theoretical insights of academia with the pragmatic needs of businesses striving to effectively incorporate sustainability into their reporting practices.

The dual focus of the book is instrumental in ensuring a well-rounded contribution to both academia and practice and positions the book as a valuable resource, not just for academics seeking theoretical depth but also for businesses navigating the practical challenges of sustainability reporting.

Within the academic sphere, this book assumes a pivotal role as a valuable resource tailored to the needs of scholars, researchers, and students. It systematically engages with the academic discourse surrounding sustainability reporting and regulation, offering a comprehensive overview of the field's evolutionary trajectory and its current status. In the corporate sphere, this book is designed to meet the varied requirements of companies in different phases of their sustainability journey. For those newly venturing into sustainability, the book serves as a fundamental guide. It not only sheds light on the reasons and advantages of undertaking sustainability reporting but also offers practical insights into crafting reports in line with global best practices. For companies already familiar with sustainability practices, the book functions as a strategic instrument. It not only keeps them updated on the latest regulatory initiatives but also positions them to effectively navigate challenges and capitalize on emerging opportunities. By staying proactive in understanding the dynamic regulatory landscape, these companies are not just prepared to face future challenges but also to identify and leverage existing opportunities in the dynamic field of sustainability reporting.

REFERENCES

Aureli, S., Magnaghi, E., & Salvatori, F. (2019). The role of existing regulation and discretion in harmonising non-financial disclosure. *Accounting in Europe, 16*(3), 290–312.

Bebbington, J., Kirk, E., & Larrinaga, C. (2012). The production of normativity: A comparison of reporting regimes in Spain and the UK. *Accounting, Organizations and Society, 37*(2), 78–94.

Bold, F. (2017). Compliance and reporting under the EU non-financial reporting directive: Requirements and opportunities, Czech Republic, Brussels, Brno.

Clarkson, P. M., Li, Y., Richardson, G. D., & Vasvari, F. P. (2008). Revisiting the relation between environmental performance and environmental disclosure: An empirical analysis. *Accounting, Organizations and Society, 33*(4–5), 303–327.

CSRD. (2022). Directive 2022/2464/EU of the European Parliament and of the Council of 14 December 2022 amending Regulation (EU) No 537/2014, Directive 2004/109/EC, Directive 2006/43/EC and Directive 2013/34/EU, as regards corporate sustainability reporting.

Deegan, C. and Unerman, J. eds., (2011). Financial Accounting Theory. 2nd ed. Berkshire: McGraw-Hill.

Delbard, O. (2008). CSR legislation in France and the European regulatory paradox: An analysis of EU CSR policy and sustainability reporting practice. *Corporate Governance, 8*, 397–405.

Doshi, A. R., Dowell, G. W., & Toffel, M. W. (2013). How firms respond to mandatory information disclosure. *Strategic Management Journal, 34*(10), 1209–1231.

KPMG. (2022). *KPMG global survey of sustainability reporting 2022.* https://kpmg.com/it/it/home/insights/2022/12/kpmg-global-survey-of-sustainability-reporting-2022.html

La Torre, M., Sabelfeld, S., Blomkvist, M., Tarquinio, L., & Dumay, J. (2018). Harmonising non-financial reporting regulation in Europe: Practical forces and projections for future research. *Meditari Accountancy Research, 26*(4), 598–621.

Luque-Vilchez, M., & Larrinaga, C. (2016). Reporting models do not translate well: Failing to regulate CSR reporting in Spain. *Social and Environmental Accountability Journal, 36*(1), 56–75.

Martin-Sardesai, A., & Guthrie, J. (2019). Social report innovation: Evidence from a major Italian bank 2007–2012. *Meditari Accountancy Research, 28*(1), 72–88.

NFRD. (2014). Directive 2014/95/EU of the European Parliament and of the Council, 22 October 2014, amending Directive 2013/34/EU as regards disclosure of non-financial and diversity information by certain large undertakings and groups.

O'Donovan, G. (2002). Environmental disclosures in the annual report. *Accounting, Auditing & Accountability Journal, 15*(3), 344–371.

CHAPTER 2

The Process of Sustainability Reporting

Abstract This chapter presents the sustainability reporting process drawing on the hierarchical staged process model. According to this model, the sustainability reporting process include five stages, which involve decisions related to the following issues: 1) Why does the entity disclose sustainability information? 2) Who are the stakeholders to whom the sustainability report is directed? 3) What information and issues should be included in the sustainability report? 4) How should the sustainability report be prepared? 5) Should the sustainability report be assured?

Keywords Sustainability reporting process · Motivation · Stakeholders · Materiality · Assurance

2.1 THE HIERARCHICAL STAGED PROCESS MODEL

Sustainability reporting is a structured process undertaken by organizations to transparently communicate economic, environmental, and social-related information to stakeholders. This process typically begins with the identification of relevant sustainability issues through stakeholder engagement and materiality assessments. Organizations then collect and analyze data related to these issues, following established reporting frameworks and standards. The next step involves the preparation and publication of a sustainability report, which outlines the organization's performance, goals, initiatives, and challenges in the realms of sustainability. This report serves as a means of accountability and helps stakeholders, including investors, consumers, employees, and communities, make informed decisions about the organization's activities and contributions to sustainability. Finally, the organization may undergo external assurance to validate the accuracy and reliability of the reported information, ensuring credibility and trustworthiness in the eyes of stakeholders.

Although there is no universal reporting process applicable to all companies, several commentators have characterized sustainability reporting as a hierarchical staged process, whereby the decisions taken at each stage in the hierarchy determine the issue to be considered and decided in the subsequent stages (Rinaldi et al., 2014). Deegan and Unerman (2011) propose that there are five steps involved in the sustainability reporting process, which require decisions related to the following issues:

1. *Why*: Why does the organization report? What are the motivations for providing sustainability information?
2. *Who*: To whom is the reported information directed? Who are the stakeholders addressed in the report?
3. *For what*: What sustainability information is reported on? What issues are material for the organization, as well as what information do the targeted stakeholders need and are interested in?
4. *How*: How is the report compiled? In what form and format is the information disclosed and communicated?
5. *Audit*: Is the report assured?

As sustainability reporting has historically been voluntary, organizations have had the autonomy to determine not only whether to engage in reporting but also which topics to disclose, the types of information and metrics to provide, and the format to use. However, this flexibility has resulted in a challenge for comparability across reports. The emergence of various reporting standards as well as the increase in mandatory sustainability reporting requirements (as discussed in Chapter 3 of this book) is beginning to address this issue by providing guidance and frameworks for consistent reporting practices. Thus, the application of one reporting framework over another, or the presence of mandatory sustainability reporting requirements, affects the structure of the entire sustainability report, particularly in terms of audience, content, and format.

2.2 MOTIVATION FOR SUSTAINABILITY REPORTING: STRATEGIC VS. HOLISTIC

The first issue in the sustainability reporting process concerns the philosophical motivations that drive managers to report sustainability information, in situations where this reporting is not mandatory. Despite there being a combination of multiple motivations guiding any organization's sustainability reporting, literature places these motives on a continuum between purely economic motives on one end and ethical motives on the other (Deegan and Unerman, 2011). Motivations toward the purely economic end of the continuum tend to be classified in the literature as *strategic*, whereas motivations that incorporate a reasonable amount of ethical content tend to be classified as *holistic*.

The holistic perspective regards sustainability reporting as a process to make business practices more socially and environmentally sustainable. This perspective is based on the assumption that companies have the responsibility (mainly ethical in nature) to care about their social and environmental impacts and contribute to sustainable development. From this perspective, sustainability reporting should be aimed at holding organizations accountable for all their significant economic, social, and environmental impacts on all stakeholders, irrespective of their financial relevance for the organization. The strategic perspective highlights that sustainability reports can be strategically used as a marketing tool, to contribute to the achievement of an organizations' goals. Then, sustainability reporting is regarded as driven by economic and profit motives, building on the assumption that the company has a single responsibility: to create economic value for its shareholders. Various theoretical perspectives highlight how sustainability reporting can help companies to maximize their profits, by improving the relationships and support of economically powerful stakeholders, gaining and maintaining legitimacy, or improving corporate reputation.

The motivations driving companies to engage in sustainability reporting may influence the choice to apply one reporting framework over another. Companies that engage in sustainability reporting for ethical motives are more likely to adopt the GRI Standards that are aimed at enabling companies to be transparent about the most significant social and environmental impacts on stakeholders. Conversely, companies disclosing on social and environmental issues driven by economic motives are more likely to adopt other standards (e.g., SASB, IIRC, CDSB, TCFD, and ISSB) that provide a framework for disclosing on the financial risks and opportunities related to sustainability issues.

2.3   THE AUDIENCE: STAKEHOLDER AND STAKEHOLDER ENGAGEMENT

Once established the philosophical motives to engage in sustainability reporting, the following stage in the reporting process involves identifying the users to whom sustainability disclosures will be directed. In general terms, sustainability reporting addresses the information needs and expectations of organizational stakeholders. Stakeholder is a general term, which commonly refers to "any groups or individuals who can affect or are affected by an organisation's activities" (Freeman, 1984). Then, stakeholders are persons or groups that have, or claim, ownership, rights, or interests in a corporation and its activities, past, present, or future. Such claimed rights or interests are the result of transactions with, or actions taken by, the corporation, and may be legal or moral, individual or collective.

Stakeholders are usually classified into primary and secondary stakeholders (Clarkson, 1995). Primary stakeholders refer to those stakeholders which engage in formal contractual relationships with a company, and without whose continuing participation the corporation cannot survive as a going concern. Primary stakeholders typically include shareholders and investors, employees, customers, suppliers, as well as the governments and communities that provide infrastructures and markets. There is a high level of interdependence between the corporation and its primary stakeholder group: it is widely recognized that companies cannot survive without the consent of these primary stakeholders and consequently should pay attention to create sufficient wealth, value, or satisfaction. Secondary stakeholder groups are defined as those who influence or affect, or are influenced or affected by, the corporation, but they are not engaged in transactions with the corporation and are not essential for its survival. Secondary stakeholders include the media, community groups, religious groups, and other non-governmental organizations (NGOs). Increasing evidence documents that also secondary stakeholders are able to induce companies to respond to their needs, as they have the capacity to mobilize public opinion in favor of, or in opposition to, a corporation.

The identification of stakeholders to whom the organization needs to report has to take place after the motives for sustainability reporting have been determined, because the range of stakeholders to be considered by any organization will be directly dependent upon its motives for engaging in sustainability reporting.

Corporations engaging in sustainability reporting for strategic motives will tend to focus on the expectations and information needs of those stakeholders who are able to exert the greatest economic influence on the organization's operations. The most economic powerful stakeholders vary depending on the context and the sector in which the company operates, but they typically include investors, consumers, and employees. Conversely, companies whose motives for engaging in sustainability reporting are grounded in the holistic perspective are more likely to address the information needs of a broader range of stakeholders, including those who may not directly influence the organization's economic success, in order to be responsible and accountable to all those upon whom the organization might have social and environmental impacts.

The heterogeneity of audiences to whom sustainability reports may be directed is also visible within the sustainability reporting standards. While the GRI Standards provide guidance to enable organizations to disclose sustainability information to a wide range of stakeholders, other prominent standards (e.g., SASB, IIRC, CDSB, TCFD, and ISSB) consider investors as the primary targets of sustainability information. For instance, the objective of ISSB's IFRS S1 is to require an entity to disclose information about its sustainability-related risks and opportunities that is useful "to primary users of general purpose financial reports in making decisions relating to providing resources to the entity" (ISSB, 2023, p. 6).

2.4   THE CONTENT OF SUSTAINABILITY REPORTING

The third stage of the sustainability reporting process concerns what type of issues and information to be disclosed in the report. The process through which organizations identify and select sustainability issues and information to be disclosed in a report is the materiality assessment process (materiality is discussed in Chapter 5 of this book).

Materiality serves as the filter for selecting the information to include in the report, based on its relevance to users. The key point is that the sustainability report should only contain material information, which is information useful to the stakeholders to which the report is directed.

A crucial element of the materiality analysis is engagement and dialogue with stakeholders, since identifying the expectations and the information needs of stakeholders is crucial to target the sustainability report toward meeting these needs. However, there is no single approach to materiality in sustainability reporting, since the criteria for evaluating which information is relevant to users may vary. There are two main approaches to materiality: impact and financial materiality.

When there are no mandatory requirements, the selected approach to materiality is dependent upon the motives driving companies to engage in sustainability reporting, as well the standards and frameworks adopted. Companies in sustainability reporting for holistic motives are more likely to adopt the impact materiality approach, which requires including in the report those sustainability issues on which the company generates the most significant (positive or negative) impacts. Impact materiality is adopted by the GRI Standards. Conversely, companies engaging in sustainability reporting for economic and financial motives are more likely to adopt financial materiality, which require to disclose on those sustainability issues that generate the most significant risks and opportunities on the companies, and are therefore likely to influence its financial value. For instance, financial materiality is adopted by SASB and ISSB.

A third perspective on materiality, known as *double materiality*, merges the two, as it requires considering as material those issues that are material from the perspective of impacts or from the financial perspective. This perspective is used in the EU's CSRD.

2.5   FORM AND FORMAT OF THE REPORT

The fourth stage of the reporting process involves selecting the appropriate form and format for sustainability reporting. Organizations have various options for reporting channels.

Corporate sustainability disclosure typically manifests within corporate reports, either as part of annual reports or as separate standalone sustainability reports, or integrated reports.

According to findings from IFAC (2024), there is a lack of convergence in the types of reports used for sustainability reporting across the globe. For instance, in 2022, the vast majority of US (93%) and Chinese (82%) companies utilize a sustainability report. In contrast, UK (91%) and German (73%) companies primarily incorporate sustainability information within their annual reports, while integrated reporting is more prevalent in South Africa (100%), Japan (85%), and France (64%).

The concept of communication channels extends beyond corporate reports. Sustainability information can also be disseminated through press releases, particularly to report on specific events or address specific stakeholders. Additionally, corporate spokespersons may communicate sustainability information through speeches and presentations at private or public events. Moreover, sustainability information has become increasingly accessible through corporate websites and social media channels, allowing companies to reach and engage with a wider audience. Similar to the earlier stages, the choice of reporting channels depends on the motivations behind a company's engagement in sustainability reporting. Companies motivated by strategic objectives are more likely to concentrate on a select few reporting channels preferred by their primary economic stakeholders. Conversely, companies with broader motivations tend to utilize a wider array of communication channels to engage with a more diverse range of stakeholder groups.

2.6   SUSTAINABILITY REPORTING ASSURANCE

The primary source of criticism, skepticism, and perplexity surrounding sustainability reports lies in the perceived reliability and credibility of these documents in accurately portraying companies' actual sustainability performance. In response to these concerns, a key solution to enhance the credibility of sustainability reports has emerged: the introduction of external verification, commonly referred to as sustainability reporting assurance.

Assurance refers to "the methods and processes employed by an assurer to assess an organisation's disclosures about its performance as well as underlying information, processes, and systems, using suitable criteria and standards in order to increase credibility. Assurance includes the communication of the results of the assurance process in an Assurance Statement" (Accountability, 2020, p. 36). This process involves independent audits or reviews conducted by an independent organization (i.e., assurance provider) to validate the accuracy, completeness, and reliability of the information disclosed in the reports.

The decision to assure a report is at the discretion of the reporting organization, although there are some jurisdictions where assurance is subject to mandatory regulation (as will be discussed in Chapter 3 of this book). Nevertheless, the use of assurance for sustainability reporting has significantly increased, with around 63% of corporations that report on sustainability matters included in the Global250 list having their report assured in 2022, compared to 46% in 2011 and 30% in 2005 (KPMG, 2023). Similarly, findings from the IFAC suggest that the rate of sustainability assurance continued to steadily progress across most jurisdictions from 51% in 2019 to 69% in 2022 (IFAC, 2024).

By undergoing external verification, companies can provide stakeholders with greater confidence in the integrity and transparency of their sustainability disclosures, thereby addressing criticisms and fostering trust in their sustainability efforts. Sustainability reporting assurance services are offered by both the most important accountancy firms and smaller non-statutory audit providers, including consultancy firms and NGOs. Although statutory audit firms continue to perform the majority of sustainability assurance globally, with a share of 73% in the year 2022, non-EU jurisdictions are less likely to use audit firms for assurance (IFAC, 2024).

As the demand for sustainability assurance services increased, different standards began to emerge. The two most popular standards used in sustainability assurance are the *International Standard on Assurance Engagement 3000* (ISAE 3000) developed by the International Auditing and Assurance Standards Board (IAASB) and the *AA1000 Assurance Standard* (AA1000AS) developed by Accountability, a UK-based consultancy firm.

ISAE 3000-standard is used to provide assurance over non-financial information. The first version of ISAE 3000 was issued in 2003 with a revision in 2013. The standard can cover various subject matters ranging from sustainability or governance topics to information security. ISAE 3000 recognizes two types of report: "Type 1" report provides assurance on design and implementation of controls on a certain date, while "Type 2" report provides assurance on design, implementation, and continuous effectiveness of controls during certain time period, usually one year. Reasonable assurance represents a substantial level of confidence, although not absolute, where the auditor asserts the material correctness of the reported information. Conversely, limited assurance indicates that the auditor has not identified any significant modifications that are necessary. Limited assurance places greater reliance on representations provided by the company's management team as a source of information. This approach involves less thorough verification of source documents compared to engagements with reasonable assurance, a less comprehensive understanding of processes and controls, and a reduced level of scrutiny regarding the inclusion of data and topics in the report. Presently, the majority of companies engaging an audit firm for reviewing their ESG reporting opt for limited assurance (IFAC, 2024).

The AA1000AS v3 is designed to be a globally leading benchmark for external assurance of organizations' adherence to their stated sustainability goals and principles. It provides important guidance on the methods and processes that should be employed by independent, external assurance providers to maximize the credibility of their findings when assessing organizations' disclosures about their sustainability performance and underlying sustainability-related information, processes, and systems, as well as for issuing those findings in a formal Assurance Statement. The AA1000AS v3 describes: (a) How to define the scope and preconditions to be met when accepting an assurance engagement where the standard is used; (b) how to perform an engagement in accordance with the standard; (c) how to issue the final Assurance Statement and optional Report to Management. There are two types of AA1000AS v3 assurance engagement: Type 1 and Type 2. For a Type 1 assurance engagement, the assurance provider shall review and assess the extent of the organization's adherence to all four AA1000 Accountability Principles (inclusivity, materiality, responsiveness, and impact) and provide relevant findings and conclusions. For a Type 2 assurance engagement, the assurance provider shall assess the extent of the organization's adherence to all four AA1000 Accountability Principles (inclusivity, materiality, responsiveness, and impact) and provide relevant findings and conclusions, and shall, additionally, assess and evidence the reliability and quality of specified sustainability performance and disclosed information, providing relevant findings and conclusions.

REFERENCES

Freeman, R. E. (1984). *Strategic management: A stakeholder approach.* Cambridge University Press.

International Sustainability Standards Board (ISSB). (2023). *IFRS S1 general requirements for disclosure of sustainability-related financial information.* https://www.ifrs.org/issued-standards/ifrs-sustainability-standards-navigator/ifrs-s1-general-requirements/#about

International Evolution of Non-financial Disclosure and Sustainability Reporting

Abstract Traditionally, the field of non-financial disclosure was characterized by a voluntary approach, based on the lack of specific disclosure obligations imposed on companies. However, reporting sustainability issues has become increasingly important over the years. Companies began to provide socio-environmental information on a voluntary basis (i.e., in the absence of specific laws or provisions) in order (for instance) to enhance their image, reputation, trust, and consensus with stakeholders and to meet their growing demand for non-financial disclosure. The voluntary initiatives clearly assumed that financial information was not sufficient to accurately delineate a firm's total value and understand its future developments. However, the content of voluntary reports was often very variegated and the information disclosed in them was difficult to compare. The voluntary approach initially embraced by the European Union was followed by several regulatory interventions on non-financial disclosure. The chapter examined such European transition from voluntary to mandatory non-financial disclosure, highlighting also the role of international reporting standards and presenting several international cases that appear significant for the adopted approaches concerning sustainability disclosure and reporting.

Keywords Non-financial disclosure · Corporate Sustainability Reporting Directive · European Sustainability Reporting Standards

© The Author(s), under exclusive license to Springer Nature
Switzerland AG 2024
C. Mio et al., *Sustainability Reporting*, Palgrave Studies in Impact
Finance, https://doi.org/10.1007/978-3-031-58449-7_3

3.1   Origins, Drivers, and Initial Development of Non-financial Disclosure and Sustainability Reporting

Traditionally, the field of non-financial disclosure was characterized by a voluntary approach, based on the lack of specific disclosure obligations imposed on companies. However, reporting sustainability issues has become increasingly important over the years (Deegan, 2017; Haller et al., 2017). Companies began to provide socio-environmental information on a voluntary basis—i.e., in the absence of specific laws or provisions—in order (for instance) to enhance their image, reputation, trust, and consensus with stakeholders and to meet their growing demand for non-financial disclosure. Indeed, from the end of the twentieth century (the first attempts were recorded in the 1970s; Hahn & Kühnen, 2013; Mathews, 1997), socio-environmental reporting was characterized by the disclosure of voluntary communication models, such as environmental, social, or sustainability statements or budgets. However, the drafting of reports, in which NFI is described according to the voluntary initiative of companies, meant that the content of these documents was often very variegated and the information disclosed in them was difficult to compare. Indeed, no specific drafting rules were laid down and no specific organization was in charge of controlling the content of these reports. The purpose of the NFI reported by the companies was to communicate to the (outside) world the results achieved in order to improve the corporate image perceived by the community within which the company was placed, without the provision of any minimum content or any obligation to draw up.

Until the 2000s, no European legislator had attempted to regulate the non-financial reporting of companies. Only in France, in 1977, Law No. 769 (Law on Social Reporting, Law No. 77–769 of July 12, 1977. Official Journal: 1977, July 13) had obliged large companies to include certain information of a non-economic nature in their annual accounts, specifically relating to relations with employees, trade unions, and community. However, the case remained isolated as French companies regarded the requirement as a mere administrative obligation, not appreciating its real value. On the one hand, it well underlined the need for accountability to be accompanied by a full understanding of its utility for both the company and the community according to a logic of full (or double) materiality (see Chapter 4 for a focus on materiality). On the other hand, however, the need to provide a common structure and guidelines for companies' non-financial reporting began to be perceived.

The voluntary initiatives clearly assumed that financial information was not sufficient to accurately delineate a firm's total value and understand its future developments. As emphasized in the introduction, financial reporting's conceptual framework states that its objective is to provide users with relevant data to help them make investment decisions. It is strictly connected with the study on the value relevance of accounting information. It involves an in-depth understanding of value, both effective and potential, both created and destroyed, from both a financial and non-financial point of view. For this reason, both academics and practitioners have pointed out the shortcomings of a traditional short-term approach that favors shareholder value maximization as the primary and exclusive objective of each business. They have expressly claimed a reflection of the tenets that underpin reporting practices and the corporate ultimate objective, which is to generate value among stakeholders to promote sustainable development. Over time, therefore, there has been a broadening of the range of companies that have decided, of their voluntary initiative, to describe their social and environmental achievements in purposefully drafted documents. Nonetheless, the dissemination of sustainability reports is still going extremely slowly, primarily due to the companies' perception that collecting and communicating NFI involves an excessive expenditure of corporate resources. The costs incurred are not always covered by the benefits and relevance that stakeholders might attribute to sustainability reporting. Among the critical issues that are frequently raised is also the lack of reporting standards that are suitable for all companies and thus the information that is reported is often difficult to compare. Over the past decades, to increase the comparability, verifiability, and reliability of information, several (voluntary) guidelines and standards have been developed for reporting the achievements in corporate social responsibility, as the integration of sustainability issues is increasingly becoming an essential component of investment decision-making. Companies operate from a sustainable perspective by trying to anticipate and manage current and future economic, social, and environmental opportunities and risks, to create long-term value. They are worldwide being asked to produce high-quality, globally comparable information on sustainability risks and opportunities. Sustainability can be investigated by paying attention, therefore, to environmental, social, governance, and economic issues. With reference to environmental disclosure, the impact of the company's activities on the external environment and the correct use of natural resources, especially in the production processes, must be taken into consideration. From a social point of view, issues such as occupational health and safety, the protection of workers' rights and human rights, fair remuneration, and the guarantee of fair working conditions must be assessed. Governance disclosure covers issues such as ownership structure, its composition and remuneration, ethical practices, and fiscal transparency. These non-financial items can be evaluated according to an economic-financial point of view, leading to the creation of long-term value for all stakeholders. This is beneficial for shareholders, but also for the community. It may generate an increase in company profits and consequently in the remuneration of employees, and investors allowing for an increase in the resources dedicated to research and development, the improvement of corporate organization, and the enhancement of human capital. Proper attention to the issue of sustainability, therefore, may entail a series of benefits for the organization, among which can be identified a greater stimulus to innovation, the recognition of the company in the industry in which it operates, a better relationship with the community and public institutions, a better reputation, and last but not least, greater attractiveness to investors.

Indeed, among the matters that corporate stakeholders take most notice of in their decision-making today is the company's ability to produce long-term value also from a non-financial point of view. For this reason, the stakeholder-oriented approach, which aims to consider the needs and requirements of all the company's stakeholders, has become widespread in the definition of corporate strategy and objectives. A company's sustainability-oriented approach depends on the sustainability of the relationships and dialogue it manages to develop with its various stakeholders. The need to consider the company's stakeholders (and not just shareholders) has led to a redefinition of the relationship between company and social value, seeking to strike a balance between economic growth, environmental protection, and social justice. With the publication of the report "Who Cares Wins" by the United Nations World Pact in 2004, CSR was joined by the acronym ESG, consisting of the words environmental, social, and governance, which aim to indicate the three important facets that define the concept of sustainability. The relevance of variables that are not only economic in nature and the community's attention to social and environmental issues have led companies to realize that these issues, like the financial ones, must also be adequately measured and reported in annual statements to be published. In recent decades, regulatory bodies themselves have begun to exert pressure to bring companies to operate according to a sustainable approach and to adequately monitor this direction.

3.2 THE EUROPEAN TRANSITION FROM VOLUNTARY TO MANDATORY DISCLOSURE

The voluntary approach initially embraced by the European Union (EU) was followed by several regulatory interventions on NFI disclosure, the first of which dates back to **1992**. Indeed, in 1992, the European Commission, in response to the demands of the United Nations summit in Rio, presented the fifth action program with the environment and sustainable development as its main focus. It was entitled "Towards sustainability". The objective was to outline a path for the entire society and to provide the tools to achieve and maintain sustainable development. This would be achieved, first and foremost, through the increased awareness of businesses, but also through the purchasing choices of informed consumers and the investment decisions of financial institutions (Official Journal of the European Communities, C 138, 17 May 1993). The program emphasized the importance for businesses to prepare reporting also based on environmental objectives. It presupposed a commitment and involvement of society as a whole, including citizens and companies. There emerged a need for approaches and instruments, as well as financial and market-related indicators, capable of providing information related to corporate choices and impacts affecting the environment. For this reason, initiatives were also proposed in the area of accounting, with regard to possible ways of recording the financial consequences of environmental issues, to promote corporate awareness of their socio-environmental impact (2001/453/EC European Commission Recommendation on recognition, measurement, and disclosure of environmental issues in the annual accounts and annual reports of EU companies). One of the main objectives of the program was to set up mechanisms that would allow environmental accountability to be carried out. For instance, polluting waste was created by most of businesses but, usually, this cost was not included in the price of the goods or services offered. In the long term, this situation would not have been sustainable (Official Journal of the European Communities, C 138, 17 May 1993). According to the European Commission, progress at the end of the work done on the fifth program was very limited. Nevertheless, during the five years of the program, there was a growing awareness of the relevance of integrating environmental objectives and indicators into more traditional and established policies and approaches. Furthermore, the importance of such information was highlighted as a means to enable responsible choices to be made based on the consequences these would have from a socio-environmental perspective.

What has been described so far had its main focus on the protection of the environment for sustainable development and considered the reporting of non-financial issues as one among many useful tools for this purpose. The focus on the environmental dimension was also reaffirmed by the abovementioned Recommendation 2001/453/EC concerning the environmental information to be disclosed in financial statements. In this document, the European Commission considered, first of all, two main interests. The first related to investors, who needed to be aware of how companies handled environmental challenges and issues. The second concerned the authorities, who were keen to monitor the proper application of environmental regulations and the associated burdens. Although some companies voluntarily published their environmental data, there was little evidence of a corporate attitude toward this practice, especially by those organizations engaged in sectors that were particularly damaging to the environment. A critical item concerned with the costs involved in collecting and disclosing such data was highlighted. Furthermore, the difficulty in comparing the environmental information provided was denoted due to the absence of homogeneous provisions regulating it. The objective of Recommendation 2001/453/EC was clearly to stimulate greater volume, quality, and transparency of environmental information to be recorded, evaluated, and disclosed in annual reports, also considering the increasing attention shown by the entire society toward environmental issues. The reported information should have allowed an understanding of the mutual impact between the development of the business and environmental issues, and the corporate level of environmental efficiency. It was recommended to make explicit the strategies adopted for environmental protection, especially regarding pollution, the improvements and progress made in this respect, the degree of implementation of actions, and the measures that would be taken to comply with current, or forthcoming, legislation. However, being a recommendation, it could not contain any actual constraints on transposition in the EU Member States. Indeed, the European Commission made it clear that Recommendation 2001/

453/EC was only intended to set out guidelines about environmental data on compliance with the Directives 78/660/EEC and 83/349/EEC, without precluding the possibility of alternative accounting approaches.

A further step forward in environmental reporting was the 2001 Green Paper entitled "Promoting a European Framework for Corporate Social Responsibility" (COM(2001)366). It was less dedicated to accounting and more focused on (more general) corporate social responsibility (CSR). With this measure, the European Commission wanted to initiate a debate on how CSR could be promoted. CSR refers to the process of voluntarily taking steps to participate in the progress of society and to protect the environment, integrating environmental, social, and economic issues. Consequently, this responsibility is exercised toward all stakeholders who have an interest in the company and who can influence its performance. It implies a vision in which economic growth and socio-environmental protection are integrated and coexist. One of the methods identified in the COM(2001)366 to promote this responsibility was to improve the transparency and reliability of the evaluation of specific corporate initiatives. Furthermore, it was noted that also SMEs should launch their initiatives as their contribution at the local level was crucial, especially for employment. The European Commission identified the transparency in disclosing corporate actions in the ecological and social spheres as necessary: the need for the integration of socio-environmental issues into company strategies and reports was reiterated. However, traditional models of organization and reporting did not appear adequate and reports on social responsibility, especially among multinationals, did not cover all the necessary issues. The framework and items developed within the COM(2001)366 can be considered as a kind of inspiration for future regulations concerning non-financial reporting, as will be seen later in this chapter. One of the questions posed at the end of the COM(2001)366 concerned possible actions to encourage corporate social responsibility. One of the solutions put forward was to introduce rules for the proper disclosure, comparison, analysis, and monitoring of CSR information in reporting and accounting systems. What was stated in the COM(2001)366 was then taken up in EU Communication 347/2002/EC, its follow-up, which proposed to initiate and describe a strategy to foster CSR at the EU level. There was a greater awareness that social responsibility was not just additional to corporate activity, but had to be integrated into management itself and reported on. Among the obstacles to the dissemination of socially responsible behavior, the European Commission identified the lack of transparency due to the absence of common tools provided for reporting and describing policies in this area. However, in line with the voluntary approach adopted, the European Commission emphasized that the implementation of socially responsible actions was the responsibility of the companies themselves. Governments and public institutions were recommended to encourage these behaviors to ensure a benefit to society. The European Commission recommended that listed companies with at least 500 employees describe their three-pronged approach (economic, social, and environmental) in their annual reports (COM(2002)347, Communication from the Commission concerning Corporate Social Responsibility: A business contribution to Sustainable Development). In summary, all the examined initial measures (i.e., recommendations and communications, without any binding effect) were aimed at encouraging companies to voluntarily disclose NFI going beyond the boundaries of mere economic and financial issues. In this way, several companies started to become aware of the importance of NFI and began to implement voluntary models of socio-environmental reporting. Despite this, corporate disclosure has continued to focus mainly on economic and financial information, while NFI has typically played a marginal role in most corporate reports. As emphasized by the literature, although the EU has continued to promote NFI communication, its interventions have never achieved satisfactory results in this regard. As demonstrated in the following chapter, a stream of literature has related this observation to a lack of specific disclosure obligations leading to significant changes in corporate disclosure. For this reason, **Directive 2003/51/EC** represented an important first step toward the formal recognition of ESG (environmental, social, and governance) issues. It amended Directives 78/660/EEC (concerning the preparation of the annual accounts of companies), 83/349/EEC (concerning the preparation of consolidated accounts), 86/635/EEC (concerning the preparation of the annual and consolidated accounts of banks and other consolidated financial institutions), and 91/674/EEC (concerning the preparation of the annual and consolidated accounts of insurance undertakings). It was named the "Accounts Modernisation Directive" to highlight the European Community's intention to eliminate discrepancies between different accounting directives and to harmonize companies' accounting practices as far as possible. It is considered the first intervention to normatively regulate the disclosure of information relating to the environment and employees. The main items introduced by Directive 2003/51/EC concern the management report and the consolidated management report, which are traditional documents of corporate financial reporting, and, thanks to this directive, they also become central to the disclosure of information on ESG issues. In particular, the management report is the document drawn up by the corporate administrative body to inform shareholders, institutional investors, creditors, and (more generally) stakeholders about the management activity carried out by the company and its directors. The choice of Directive 2003/51/EC to focus on the management report derives from the observation of proactive, diversified, and variable corporate behaviors that had anticipated the transposition of mandatory regulations through the voluntary presentation of NFI. The decision to include the disclosure of ESG issues in such a document (traditionally focused on the management perspective) highlighted the awareness and need for responsibility on the part of the company's management bodies. According to the amendments introduced by Directive 2003/51/EC, the ordinary and consolidated management report had to contain a faithful account of the company's performance and results of operations, consistent with the size and complexity of the business, and a description of the principal risks and uncertainties faced. The management report was a document addressed to external users aimed essentially at completing and supplementing the information in the financial statements to provide a correct and complete view of the company's situation, also giving prominence, where deemed necessary, to data that could not be deduced from the general accounts. The application of the Directive 2003/51/EC led to the coexistence of the use of both "financial indicators" (derived from the financial statements) and "non-financial indicators" (originating from information sources outside the financial statements) significant according to the company's specific activity, including information relating to the environment and employees. Specifically, it requires to report financial indicators and, only where appropriate, non-financial indicators. The statement "if applicable" concerning "non-financial indicators" imposed, first of all, a distinction between information that had to be disclosed in the management report on a mandatory basis, from other information that had to be provided only optionally and when certain prerequisites were met. This implied that the reference to non-financial indicators was necessary to be implemented only in situations where neither the financial statements nor the financial indicators were capable of meaningfully and clearly expressing the company's situation and performance. In this context, the term "non-financial indicators" was understood to mean quantitative data, often of a non-monetary nature, capable of explaining the factors influencing the company's situation depending on the sector, size, and complexity of the business. In formulating these indicators, the following items should have been taken into account: market positioning, customer satisfaction efficiency of processes, and innovation. These items were considered capable of signaling, sometimes in advance of financial indicators, trends in corporate results, especially in a long-term perspective. For all these reasons, the implementation of Directive 2003/51/EC certainly represented a step forward in non-financial disclosure, although its substantial limitations and critical matters are evident today, due to an approach that was still far away from the current vision and understanding of sustainability. Specifically, Directive 2003/51/EC did not impose specific obligations and did not imply significant changes in corporate reporting (with negligible effects on both the quantity and quality of the reported information). It allowed wide discretion and flexibility about the volume and type of disclosure to be made in the annual report, as it did not provide for explicit indicators to establish the quantity and quality of NFI relating to socio-environmental issues. For these reasons, some EU Member States (e.g., Spain, France, Portugal, Denmark) independently decided to introduce special regulations to mandate the disclosure of NFI, thus anticipating the EU's adoption of a more strictly regulated approach. Indeed, the EU has taken more than ten years (following Directive 2003/51/EC) to change its approach and issue another directive about NFI. In 2014, there was a real turning point with the **Directive 2014/95/EU** (entitled "Disclosure of non-financial and diversity information", hereinafter NFRD, 2014) that marked the transition from a voluntary approach to a regulated one in the context of non-financial disclosure. It aimed to improve the transparency and accountability of large companies on non-financial issues. The NFRD (2014) set new minimum reporting standards on environmental and social issues, specifically in relation to employee management, respect for human rights, and the fight against active and passive corruption. In addition, it aimed to introduce and reinforce good practices, increase transparency in the disclosure of NFI, and boost the confidence of corporate stakeholders. In addition to the regulatory developments described so far, several other factors can be highlighted that led the EU legislator to enact the NFRD (2014), including international initiatives, increased awareness of the socio-environmental impacts of organizations, changes in consumer and investor choices, and increased demand for information and transparency with regard to corporate impacts. The legislator's choice of opting for a directive as a regulatory instrument (instead of a regulation, for example) had the advantage of allowing individual member states to adapt the content in a way that best suits each national system while retaining the virtue of providing the most internationally comparable disclosure possible. To do this, each member state transposed the directive into its law by December 2016. According to the NFRD (2014), large undertakings and groups were mandated to draft and publish non-financial statements (that can be stand-alone or included in the annual report) to communicate their commitment, policies, and performance regarding specific non-financial topics, such as environmental and climate protection, employees' management, respect for diversity, the fight against corruption, and the violation of human rights. Specifically, Article 1(1)(1) of the NFRD (2014) stated as follows: *"Large undertakings which are public-interest entities exceeding on their balance sheet dates the criterion of the average number of 500 employees during the financial year shall include in the management report a non-financial statement containing information to the extent necessary for an understanding of the undertaking's development, performance, position and impact of its activity, relating to, as a minimum, environmental, social and employee matters, respect for human rights, anti-corruption and bribery matters, including:*

(a) *a brief description of the undertaking's business model;*
(b) *a description of the policies pursued by the undertaking in relation to those matters, including due diligence processes implemented;*
(c) *the outcome of those policies;*
(d) *the principal risks related to those matters linked to the undertaking's operations including, where relevant and proportionate, its business relationships, products or services which are likely to cause adverse impacts in those areas, and how the undertaking manages those risks;*
(e) *non-financial key performance indicators relevant to the particular business"*.

The NFRD (2014) also guided companies in selecting the content to be included in the non-financial statement, allowing companies discretion and flexibility as to whether to expand the scope of the disclosure if it was relevant to a true and fair view of the corporate situation. Specifically, the contents of the disclosure covered six specific thematic areas (environmental, social, human rights, active and passive anti-corruption, and diversity issues) to ensure a full understanding of the business. If one or more of these issues were not reported, it was necessary to provide reasons for this. Indeed, the NFRD (2014) stated that companies could not provide the required disclosure as long as a clear and articulate explanation of the non-disclosure was provided and as long as the omission did not compromise a true and fair understanding of the company's performance, its results, and the impact of its activities. In two cases (i.e., omitted material information and information published after the deadline), there was also a penalty. Moreover, the possibility of providing and preparing an additional voluntary non-financial statement was recognized. The legislator in transposing the NFRD (2014) had envisaged that also entities other than those obliged could voluntarily prepare a non-financial statement to promote greater transparency about the topics listed above. In disclosing the required information, companies could rely on reporting standards and guidelines issued by authoritative supranational, international, or national bodies of a public or private nature. Companies could also adopt other methodologies, relying on some reporting standards and/or additional principles and indicators independently identified and supplementary to those provided by the reporting standards. Instead, the essential disclosure criteria to be followed in disclosing NFI were identified:

- materiality, as only truly relevant information was to be provided considering the features and activities of the company, in relation to its business and risk profile. To this end, a materiality analysis was carried out to assess the main sustainability issues;
- fairness and comprehensibility, as the disclosure had to take into account all matters, both positive and negative, evaluated and presented in an unbiased manner to guarantee the principle of transparency;
- completeness and conciseness, as the information had to be able to provide a complete picture of the impacts, results, and risks generated about material issues;
- strategic and long-term vision, as the information, while covering past events and results, also had to include future objectives and commitments. The inclusion of forward-looking information

allowed users to be able to measure the company's progress in achieving its long-term objectives;

- stakeholder orientation, as the non-financial statement, being a document complementing the financial statements, was addressed to the attention of present and potential stakeholders;
- consistency, the information presented in the non-financial statement had to be consistent with the business activity, context, and sector of the company and there had to be continuity in the reporting methodologies and standards implemented;
- comply or explain the principle, if the company decides to omit certain information, the obligation arises to indicate within the report itself the reasons for choosing not to provide such information.

The improvements of the disclosure through the non-financial statements required by the NFRD (2014) implied the orienting of the investment decisions implemented by both individuals and universal owners (large institutional investors, especially governments and management funds) toward companies proving to be more virtuous in terms of the social and environmental impacts produced and more attentive to sustainable management and production strategies. Unlike past initiatives, the Directive 2014/95/EU (NFRD, 2014) detailed the NFI-related issues that should have been disclosed, where to provide them and the responsibilities and penalties to be applied in case of non-compliance. This regulatory approach indicated that a new era in corporate disclosure had begun, in which the new reporting requirements were accompanied by additional benefits from the entire system in terms of corporate reputation and image, in terms of benefits, or other rewarding instruments. Despite the benefits, some critical points of the NFRD (2014) have emerged and led to further regulation advances.

To revise the Directive 2014/95/EU, on April 21, 2021, the European Commission adopted a proposal for a Corporate Sustainability Reporting Directive (CSRD) that was approved on November 10, 2022. Subsequently, on December 16, 2022, the text of the CSRD (**Directive 2022/2464**, hereinafter CRSD, 2022) was published in the Official Journal of the EU, replacing the term "Non-financial Reporting" with "Sustainability Reporting". Amending the requirements of the Directive 2014/95/EU (NFRD, 2014), the CSRD (2022) extends the scope of mandatory sustainability reporting to all large companies and all companies listed on regulated markets (except no-listed SMEs and all micro-enterprises), requires the audit (assurance) of reported information, introduces more detailed reporting requirements, including the provision to apply mandatory EU sustainability reporting standards (as issued by the European Financial Reporting Advisory Group—EFRAG), and requires to digitally "tag" the reported information. With the CSRD (2022), the EU intends to achieve greater transparency, comparability, and uniformity regarding NFI disclosed. This is also accompanied by a willingness to increase the quality and the real informativeness of the reporting. The CSRD (2022) is one of the cornerstones of the 2030 Agenda for Sustainable Finance and the European Green Deal, which is the EU's new growth strategy and whose objective is to transform the EU into a modern, resource-efficient, and competitive economy with net zero greenhouse gas emissions by 2050. It also aims to protect, conserve, and enhance the EU's natural capital and to protect the health and well-being of citizens from environmental risks. In doing so, it places at the center of European policy the goal of building an economy that protects people and provides stability, jobs, growth, and sustainable investments. These objectives are particularly important considering the socio-economic damage caused by the COVID-19 pandemic and the need for a sustainable, inclusive, and equitable recovery. For these reasons, in March 2018, the European Commission published an "Action Plan for Sustainable Finance", outlining a strategy and measures to achieve a financial system capable of promoting development that is truly sustainable in economic, social, and environmental terms. To implement the Paris Agreement on climate change and the 2030 Agenda, the European Commission has set the following objectives: direct capital flows toward sustainable investments for sustainable and inclusive growth; manage the risks arising from climate change (resource depletion, environmental degradation, and social issues); promote transparency and long-term vision in financial and business activity. From this perspective, corporate disclosure of reliable and comparable sustainability information is a prerequisite for achieving these goals. Furthermore, by publishing the information required by CSRD (2022), certain advantages and benefits can be pursued, such as: increasing awareness and understanding of risks and opportunities and improving the dialogue with all stakeholders. The EU adopts the stakeholder approach, without privileging any particular category of actors, but intends to promote measures and regulations to mandate the preparation of sustainability reporting. This represents the result of a complex process that requires the corporate understanding of guidelines, the identification of objectives, the development of strategies, and the implementation of activities to achieve a real and pervasive cultural change in the company. In compliance with regulatory developments and response to ever-increasing market demand, the disclosure of sustainability issues according to CSRD (2022) is published and integrated into the management report, which is included in the annual and/ or consolidated financial statements. It aims to increase transparency in environmental, social, and governance issues, to counteract greenwashing, and to strengthen the sustainable orientation of the European market by laying the foundation for sustainability reporting standards on a global level. Therefore, the CSRD takes a further and considerable step forward, putting ESG-related information on the same level as financial information and fostering greater alignment between the two, as well as extending the scope of sustainability reporting. As mentioned above, the CSRD (2022) considerably expands the number of companies that are required to disclose sustainability-related information in their management report. The scope is thus extended to approximately 49,000 European companies, compared to the previous 11,700. The directive will apply to all large companies, but also to listed SMEs. The legislator considers particularly important both that investors have access to adequate information on the sustainability of listed companies and that all listed companies may not run the risk of being excluded from investment portfolios due to a lack of ESG-disclosed information. Therefore, only unlisted SMEs and micro-enterprises are excluded from the scope of CSRD. This aims to give the European Union a frontrunner role in the context of transparency on sustainability issues. It is also realized by extending some of the new rules beyond the borders of the old continent. Indeed, the CSRD requires that third-country undertakings, that generate net revenues in the EU of more than EUR 150 million, are obliged to prepare a sustainability report. Alongside this, there is also a broadening of the "ultimate beneficiaries" of sustainability reporting that are expressly identified. Indeed, the recipients of sustainability reporting fall into three main categories: investors who wish to better understand the risks and opportunities that sustainability issues present for their investments; civil society actors, including non-governmental organizations and social partners, who expect companies to be more responsible for their impact on people and the environment; other stakeholders, who might use sustainability information to compare companies within the market. The aim is to achieve and secure a certain European harmonization of sustainability reporting. This requires the development of suitable reporting standards to overcome the current problem of comparability of reports prepared according to very different reporting standards and guidelines. Therefore, as of January1, 2024, the companies concerned must be able to fulfill the requirements of the CSRD (2022), which has delegated the European Financial Reporting Advisory Group (EFRAG) to issue and revise the European Sustainability Reporting Standards (ESRS). This is one of the main innovations of the CSRD (2022). Compared to the standards used voluntarily in the past, the ESRS broaden the reporting requirements and introduce more stringent and complex concepts to create standardized and comparable sustainability reporting and raise its quality. Therefore, information sets with a more rigid structure and mandatory minimum requirements are established, as well as audit requirements to ensure the accuracy and reliability of the reported information. Furthermore, the aforementioned digitalization of sustainability information aims to reduce the costs of collecting, reporting, consulting, comparing, and using information for companies, investors, and other (interested) stakeholders. Another important issue highlighted by CSRD concerns the "double materiality" perspective: on the one hand, companies have to explain how sustainability issues affect their business; on the other hand, they also have to report on the impact of their business on people and the environment in particular. Stakeholder surveys and benchmark analyses with industry companies, peers, and customers can be used to map material ESG issues on which the company is required to report both qualitative and quantitative information. This activity should not be limiting or incisive, but should be used as a sort of "compass": by examining the surveys and carrying out further analyses, material issues considered relevant to both the company and its stakeholders emerge. On these, the organization sets strategic priorities to draw up the resulting short- and medium- to long-term sustainability plan. This becomes a sort of public statement of corporate commitment to the realization of priority and strategic material issues. On the one hand, it addresses the mitigation of business-related risks in the ESG area; on the other hand, it maximizes the related opportunities. Furthermore, to avoid obtaining an unrealistic plan or one that is inconsistent with the company's other strategies, qualitative (with the initiatives to be undertaken) and quantitative (e.g., new certifications, net zero emission target, etc.) objectives, the relative timing (which may be short, medium, long term) and the budget for each activity should be established. Hence, the selection and quantification of targets should strike the right balance between the ambition of the sustainability plan and its feasibility, avoiding investing in targets that are only partially controllable by the company. What is envisaged in the sustainability plan should subsequently be adequately disclosed in the sustainability reporting to be included in the management report of the financial statements and to be prepared in electronic format (CSRD, 2022). For this reason, the timing of the sustainability reporting should be aligned with that of the preparation of the financial statements. The sustainability report will allow for a broader communication of medium- to long-term strategic directions, including possible business risks and opportunities, and will have to contain all the information required by the relevant standards (ESRS). The preparation of the report should follow the principles of accuracy, fairness, clarity, comparability, completeness, timeliness, and verifiability. The sustainability report should not be too long and opaque: it should be intended as a primary form of accountability and enhancement of sustainability matters. The CSRD (2022) has certainly initiated a regulatory process focused precisely on sustainability reporting, providing for a further extension of informational requirements, greater uniformity of disclosure at the international level, and additional standards and obligations also related to auditors' assurance.

3.3 INTERNATIONAL REPORTING FRAMEWORKS AND STANDARDS

The voluntary approach initially embraced by the European Union (EU) was supported by different sets of voluntary international guidelines and frameworks. Specifically, **GRI (Global Reporting Initiative)** is an independent international organization founded in 1997 whose mission is to *"help businesses and other organizations take responsibility for their impacts, by providing them with the global common language to communicate those impacts"*. GRI has several regional offices, located in Johannesburg (Africa), Singapore (ASEAN), São Paulo (Brazil), Hong Kong (Greater China Region), Bogota (Latin America), New York (North America), and New Delhi (South Asia). The organization boasts a wide network of partnerships with businesses, investors, policymakers, civil society, and labor organizations to name a few. The primary objective of these partnerships is to develop GRI Standards on sustainability reporting and encourage their adoption across the globe. GRI Standards are currently the most widely used principles in sustainability reporting and are applied by over 10,000 organizations across more than 100 countries. Indeed, since 1997 GRI has been working on defining how to prepare comprehensive non-financial reporting that can be drafted and applied by any organization. Adopting a holistic approach to accountability, GRI aims to create a global common language for organizations to report their impacts, enabling informed dialogue and decision-making around those impacts. The GRI Standards are the most internationally applied for many reasons. These include the approach focused on a wide range of stakeholders, such as employees, communities, investors, and professionals. In addition, the development and updating of the GRI Standards stem from collaborations with many governments and sustainability-oriented international organizations such as the Organisation for Economic Co-operation and Development (OECD) and the United Nations. The most recent version of the GRI Standards is the one issued in 2016 and updated in 2022 (GRI, 2023). It includes three categories of Standards: the GRI *Universal Standards*, the GRI *Sector Standards*, and the GRI *Topic Standards*. The first set of standards (called Universal) applies to all organizations and relates to general information, such as the principles adopted in reporting, contextual information about each company, and how the company deals with the issues it considers most relevant. Topic-specific standards are divided into Economic, Environmental, and Social categories and regard the information that each company should report based on its impact in such three areas. For each topic-specific standard, some ad hoc reporting suggestions and recommendations for data collection and interpretation are provided. The GRI Standards focus on the principle of materiality: the issues to be reported can influence stakeholders' decisions and are considered effectively relevant in analyzing the economic, environmental, and social impact of the business (see Chapter 5). The broad adoption of the GRI Standards has highlighted the increasing volume of provided and requested information about corporate sustainability, but it has also emphasized the heterogeneity of information provided in this regard and the difficulty of comparability both over time and between companies. To minimize risks associated with an increasing number of sustainability reports using heterogeneous conceptual bases, it becomes crucial to promote standards interoperability (EFRAG & GRI, 2023; Grewal et al., 2021). There emerges the need to support the implementation of a set of reliable and comparable measures (Pizzi et al., 2022). In this scenario, the GRI might still play an essential role in assisting the processes that contribute to reaching such standard interoperability (Pizzi et al., 2024). For these reasons, following considerable market demand, the foundation of the **International Sustainability Standards Board (ISSB)** was announced by the Trustees of the IFRS Foundation in November 2021. The ISSB considers the initiatives implemented by the most widely recognized sustainability reporting standards-setters, such as the **Global Reporting Initiative (GRI)**, the **International Integrated Reporting Framework (IIRC)**, and the **Sustainability Accounting Standards Board (SASB)**. It has international support: the G7, the G20, the **International Organization of Securities Commissions (IOSCO)**, the Financial Stability Board, African Finance Ministers, and Finance Ministers and Central Bank Governors from more than 40 jurisdictions have embraced the ISSB's aim to develop and promote sustainability disclosure standards. The ISSB builds on the work of market-led investor-focused reporting initiatives such as the **Climate Disclosure Standards Board (CDSB)**, the **Task Force for Climate-related Financial Disclosures (TCFD)**, the Value Reporting Foundation's Integrated Reporting Framework and industry-based SASB Standards, as well as the World Economic Forum's Stakeholder Capitalism Metrics. The ISSB aims to develop high-quality international standards of sustainability disclosures, addressing the information demands of investors, enabling corporations to offer full sustainability information to global capital markets, and facilitating comparability with (other types of) disclosures that are jurisdiction-specific.

In March 2022, the ISSB presented the first version of the *IFRS sustainability standards*, specifically IFRS S1 and IFRS S2. Based on the contents of these two standards, companies should report on all ESG issues, focusing on how much and how such matters impact corporate value. IFRS S1 (*General Requirements for Disclosure of Sustainability-related Financial Information*) presents the basic features of sustainability reporting, in particular, it specifies that the information provided should be complete, consistent, comparable, verifiable, and such as to highlight the impact of the disclosed matters on company value. They regard a broad range of topics, from the workforce to the consumption of natural resources, as well as the risks and opportunities arising from sustainability. The standard also emphasizes that the information to be disclosed must be material (i.e., if omitted may change the stakeholders' perception of the company's value). The information provided should therefore be complete and accurate, with particular attention also paid to issues related to corporate governance and the bodies responsible for assessing risks and opportunities linked to ESG items. While the first standard presents the basic features of sustainability disclosures, IFRS S2 (*Climate-related disclosures*) focuses on climate-related issues. The proposed second standard starts from the consideration of the risk that climate change poses to companies and the economic system. It states that companies should seize the opportunities arising from strategies aimed at climate change mitigation and adaptation. Companies should therefore inform on how their annual accounts, financial flows, and strategy are affected by climate change. Furthermore, stakeholders are increasingly interested in climate-related objectives. Companies should communicate what their goals are and how they intend to achieve them. Therefore, companies should disclose:

- their governance and procedures followed to monitor and manage climate-related risks;
- how climate change influences the company's strategy and business model;
- the risks arising from climate change and how they are managed and mitigated;
- the target parameters considered to manage and monitor climate change risks and opportunities.

The project work promoted by ISSB highlights the need for the introduction of common reporting standards for all companies to ensure comparability in the reporting of sustainability information. The elaboration and dissemination of standards valid for all companies in the European Union should on the one hand be based, as far as possible, on standards (such as those elaborated by GRI) that already exist and are used internationally (so as not to require drastic corporate changes in reporting). On the other hand, they should follow the regulatory developments highlighted in the previous section, proving to be adequate to best represent all the information required by the European legislator. The CSRD (2022) identifies the European Financial Reporting Advisory Group (EFRAG) as the appropriate body to develop these common reporting standards called European Sustainability Reporting Standards (ESRS). The first two general and transversal ESRS emphasize that the concept of double materiality underlies sustainability reporting.

This perspective takes into account both the financial and the environmental and social impact dimensions, stating that they should be assessed together (see Chapter 5). Alongside these two general standards, EFRAG has developed ten other ESRS related to ESG issues; in particular, five standards related to the environment (ESRS E 1–5), four related to social matters (ESRS S 1–4), and finally one related to governance (ESRS G). Specifically, ESRS can be outlined as follows:

- ESRS E1, *climate change*: companies are asked to specify how they influence climate change, from the points of view of both negative impact and actions taken to counter it;
- ESRS E2, *pollution*: companies are asked to clarify how they affect the pollution of air, water, and soil, as well as living organisms and food resources. Here again, companies are asked to explain how they negatively affect these aspects and what actions they take to counteract them;
- ESRS E3, *water, and marine resources*: this standard leads companies to disclose how they affect water and marine resources, specifying any actions they take to protect water, specifically referring to water consumption and discharges;
- ESRS E4, *biodiversity*: companies should specify how they affect biodiversity and ecosystems, considered as the forms and structures of life on earth;
- ESRS E5, *resource use, and circular economy*: companies should explain how they affect the use of natural resources, with a view to a regenerative consumption model, involving the recycling of materials and the use of products for as long as possible;
- ESRS S1, *workforce*: companies must describe how they impact their workforce, paying major attention to social issues, equal treatment, and privacy;
- ESRS S2, *workers in the value chain*: this standard requires each company to describe how the organization affects workers along the value chain. This standard includes all workers who, although not part of the workforce, are engaged upstream or downstream in the value chain;
- ESRS S3, *affected communities*: the company describes how its activity affects all the people residing in the geographical area surrounding the organization. It should therefore describe how local

communities are affected by the business and how the company acts to reduce negative impacts;

- ESRS S4, *consumers and end users*: the company defines how the goods and/or services it offers impact consumers and end users;
- ESRS G1, *business conduct*: this standard involves describing how the company operates, making explicit the features of its business model, and whether it is transparent and sustainable.

These ten ESRS apply to all companies required to prepare sustainability reporting according to the CSRD (2022). The setting of non-financial disclosures is gaining ground globally and developing to meet the needs of users, to ensure sustainable development, and to improve the management of risk arising from ESG issues for companies. This is not only happening at the European level, although it represents the most pioneering and forward-looking framework but changes can be observed worldwide (Harper Ho, 2020). In the international panorama, several countries have responded promptly to the new measures adopted by the EU and to the informational needs expressed by the global market, promoting specific actions on sustainability and establishing both quantitative and qualitative disclosure requirements. In the following sections, we highlight some international cases that appear significant for the approach adopted concerning sustainability disclosure and reporting.

3.4   The Peculiar Market-driven Regulatory Context in the USA

In the USA, there is no uniform direction required and/or promoted by laws and standards on ESG and sustainability issues. These seem to relate to trust principles such as loyalty and the duty to act in the best interests of customers. Respecting full and fair disclosure of all material matters, any ESG approach can be adopted as long as there is due consent from external stakeholders. Indeed, in the USA, pressure on companies is not exerted by a specific category of stakeholders, but by a plurality of people, groups, associations, and entities. The great attention traditionally paid to investors' claims and more generally to the demands of the financial market has greatly influenced the fundamental features of corporate reporting, which is still characterized by the absence of standardization and external controls in the USA. Often it takes a defensive and careful attitude toward certain issues, especially regarding environmental concerns and climate change.

As early as the 1970s, the FASB (Financial Accounting Standards Board) provided rules addressing environmental disclosure specifically concerning capitalization or expensing of environmental outlays (through three Issue Statements, FASB 1991a, 1991b, 1995) and contingent liabilities (FASB, 1975, 1976). Moreover, various studies and research conducted by prominent US associations emphasized the importance of communication and reporting on environmental issues. In detail, the American Accounting Association (AAA, 1971, 1975, 1978), the National Association of Accountants (NAA, 1974), the American Institute of Certified Public Accountants (AICPA, 1973, 1976), and the Intergovernmental Working Group of Experts on International Standards of Accounting and Reporting (ISAR) dealt with issues concerning environmental disclosure through various approaches and for different purposes, emphasizing its relevance for both internal and external users (Millstone & Watts, 1992). Indeed, numerous corporate managers were already aware of environmental concerns at that time and attempted to satisfy environmental disclosure requests voluntarily (Wiseman, 1982). According to a United Nations survey, 86% of the 222 examined companies made some sort of environmental disclosure in their 1990 annual reports (United Nations, 1992). As early as the 1990s, American literature highlighted the need to establish policy for environmental reporting: *"policy makers should require, at a minimum, a company's environmental policy and objectives, contingent and actual liabilities, current and estimated capital expenditures, asset valuation, risk assessment, performance measures, and the impact of environmental costs on earnings and earnings per share. This minimum requirement should provide stakeholders with information that will allow them to assess the cash flow effect as well as the societal impact of an entity's environmental concerns"* (Gamble et al., 1996, p. 314). This empirical study pointed out how environmental accountability practices varied greatly among and within the selected entities and were not systematically implemented over time. The sampled companies voluntarily provided more information than was required, indicating an overall intention of firms to report environmental information, yet several such companies disclosed the same environmental information in multiple places. This last remark reinforced the need to establish reporting guidelines for environmental issues.

Despite the growing relevance of non-financial disclosures for a growing number of recipients in the USA, corporate reporting choices on ESG issues remain primarily voluntary and market-driven. Climate change is a constant topic of discussion and seems likely to induce a shift in policy direction: there is a growing recognition of the importance of addressing climate change and managing climate-related financial risk. Indeed, the recognition of the ESG approach in SEC (Securities and Exchange Commission) consultations and the willingness to enforce consequent corporate behavior has been highlighted. The Inflation Reduction Act, enacted by the United States Congress in August 2022, combines the goals of cutting domestic inflation while addressing climate change. The act's declared purpose is to cut carbon emissions by approximately forty percent by 2030. It contains a mix of incentives designed to promote the massive introduction of clean energy. It represents the largest government investment in clean energy in US history with an allocation of approximately 370 billion USD to these initiatives.

Another important initiative, along the same lines of change just described, but in the area of accounting specifically, took place on March 21, 2022, when the SEC announced that it was developing a set of rules to enhance and standardize climate-related disclosures by public companies and in public offerings. The proposal is based on three main pillars of action aimed at promoting more informed finance:

- the introduction of a reporting framework that is compatible with the tax standard and includes information on the significant impacts that climate change has on corporate affairs;
- the obligation to make public the methods and metrics used to calculate the environmental impact of corporate activity, as well as those useful to determine the risks arising from climate change itself;
- the inclusion in annual reports of a certified statement about corporate greenhouse gas emissions.

The final SEC rules entitled "The Enhancement and Standardization of Climate-Related Disclosures for Investors" have been adopted by the SEC on March 2024. These rules will require public companies to provide certain climate-related information in their registration statements and annual reports. They will be phased in for all registrants with the compliance date dependent upon the status of the registrant as a large accelerated filer (LAF), an accelerated filer (AF), or non-accelerated filer (NAF), smaller reporting company (SRC), or emerging growth company (EGC), and the content of the disclosure provides several accommodations.

In particular, registrants will be required to disclose:

- climate-related risks that have had or are reasonably likely to have a material impact on the registrant's business strategy, results of operations, or financial condition, as well as the actual and potential material;
- if, as part of its strategy, a registrant has undertaken activities to mitigate or adapt to a material climate-related risk;
- any oversight by the board of directors of climate-related risks and any role by management in assessing and managing the registrant's material climate-related risks;
- information about a registrant's climate-related targets or goals, if any, that have materially affected or are reasonably likely to materially affect the registrant's business, results of operations, or financial condition.
- for large accelerated filers (LAFs) and accelerated filers (AFs) that are not otherwise exempted, information about material Scope 1 emissions and/or Scope 2 emissions;
- for those required to disclose Scope 1 and/or Scope 2 emissions, an assurance report at the limited assurance level, which, for an LAF, following an additional transition period, will be at the reasonable assurance level.

To summarize, while regulations governing mandatory financial reporting force US corporations to disclose certain environmental information, sustainability reporting in the USA is primarily market-driven. The largest public corporations make non-financial disclosures primarily through free-standing sustainability reports, corporate surveys, and direct engagement between investors and corporate management (Harper Ho, 2020). Although even in the USA there has been a growing demand for greater transparency on ESG issues from investors and other corporate stakeholders, and new regulation has been issued in this regard (SEC, 2024), there are still many open issues to be resolved that include: the overwhelming role of shareholders in influencing corporate choices and subsequent practices; skepticism about regulatory constraints especially in peculiar sectors of the US economic and political system; and the almost single-minded focus on environmental matters while neglecting other central issues of sustainability reporting.

3.5   The Case of the UK

Some studies claim that CSR was born in the Anglo-Saxon country at the time of the Industrial Revolution in the second half of the nineteenth century, even though the term was coined many decades later. Indeed, in the UK as early as the nineteenth century some pioneering entrepreneurs identified the importance of providing decent living conditions for their employees. The focus on employees was such that in the early 1900s, some entrepreneurs went so far as to set up the first pension fund for their workers—a truly revolutionary initiative for the time. Great Britain was one of the countries where corporate responsibility became established earlier than in other countries and where various initiatives were born to disseminate and develop activities related to social and environmental issues. Since then, UK government institutions have been at the forefront of disseminating CSR principles and promoting their adoption by all organizations to issue increasingly transparent annual reports. In particular, some measures even predate European regulations, including the Equal Pay Act of 1970, the Health and Safety at Work Act of 1974, the Sex and Discrimination Act of 1975, and the Race Relation Act of 1976. Under Blair's government (1997–2007), the world's first Ministry of CSR was established (in 2000). Reviewing the most relevant normative initiatives in the light of European measures and in particular Directive 2003/51/EC, the Companies Act of 2006 required all non-small companies to prepare a report, called "Business Review" and include it in the annual management report, where information on the environment, employees, and contractual agreements involving the company and the community is provided. Moreover, academic scholars, professional organizations, and other groups have encouraged larger firms to address these issues in their annual reports. For instance, as early as 1991, the Chartered Association of Certified Accountants developed an annual Environmental Reporting Award Scheme with three main objectives: showing all stakeholders and the entire business community which companies provide environmental disclosure; recognizing and promoting innovative efforts at conveying about environmental outcomes through offering a framework for the reporting; and providing all companies (also those that do not have such opportunity) to evaluate the positive effects of environmental disclosure. At the same time, the Environmental Research Group of the Institute of Chartered Accountants has emphasized in England and Wales since the 1990s that more environmental disclosure is especially necessary when environmental factors have an impact on a company's policies and operations, or when they affect the values of its assets or liabilities. They also suggest a list of matters that businesses ought to disclose as part of their yearly reporting cycle: the company's environmental strategy, its environmental goals, the main environmental effects of the business, and details on the steps the company has taken to achieve its environmental objectives. Furthermore, the finance directors of the top 100 UK corporations released a statement in the 1990s providing a framework for environmental reporting and urging all businesses to meet such minimum standards of reporting.

Today, it is estimated that the uptake of sustainability reporting at UK large companies currently exceeds 95%. The governmental commitment, together with the diffusion of a consolidated organizational culture focused on the attention and involvement of stakeholders, has made the UK one of the world leaders in the adoption of CSR policies and ESG disclosure.

3.6   Australia and New Zealand Adopting the Recommendations of the Task Force on Climate-related Financial Disclosures

Australia and New Zealand are paying close attention to sustainability reporting regulation and specifically to the recommendations of the Task Force on Climate-related Financial Disclosures (TCFD). It was instituted at the behest of the Financial Stability Board, which is a body charged with drafting and issuing recommendations/guidelines on climate risk disclosure. The Financial Stability Board's action stems from a joint request by the G20 finance ministers and central bank governors at the end of 2015 to analyze how the financial sector could report on climate issues. The set goal to be achieved in the medium to long term was to promote the transition to a low-carbon type of economy. However, this coveted transition had to be as gradual as possible. This included climate-related disclosure, which it was felt needed to be improved both quantitatively and qualitatively. Better disclosure would have made it possible to monitor potential risks more closely, to seize new opportunities related to climate change that had hitherto gone unexplored, and to steer both current and potential investors toward more informed choices and financial institutions toward granting loans and insurance cover. After extensive public consultation, the "Recommendations of the Task Force on Climate-related Financial Disclosures" were published in June 2017. These recommendations considered non-financial climate-related disclosures for a wide range of organizations differing by sector and jurisdiction. In particular, it was intended that the information could be included in financial reporting and be easily adaptable to all types of organizations. In addition, they had to appear particularly effective in giving useful indications of future implications concerning the financial situation and pay particular attention to risks and opportunities in a transition toward reduced carbon emissions. Climate-related risks and opportunities have been defined, specifying how these should be accounted for in annual reports. The TCFD highlighted that such financial climate-related impacts clearly result in concrete effects on the financial statements. With regard to costs, risks may have a significant impact depending on the cost structure of the company and its operational flexibility. The required cost of capital may also be lower than that of an organization that does not pay attention to climate issues at all. On the revenue side, a decrease could occur due to lower unit prices of products and services, loss of market share compared to more environmentally aware competitors, or production stoppages due to, for example, natural disasters. As far as financial repercussions are concerned, much depends on the specific organizational structure. Obsolete plants and production processes that are inadequate concerning new environmental demands and not resilient to possible climate risks will result in write-downs and subsequently in a lower value of assets and thus in a poorer ability to attract capital and financing on favorable terms. It should not be forgotten, however, that adapting requires new investments. Moreover, the financial impacts outlined so far are not always clear in both their identification and evaluation. Not realizing the possible opportunities or failing to control the risks could lead to serious consequences. Hence, the resulting disclosure is often complex and challenging. This is for several reasons: little or no knowledge within the company of climate-related issues; difficulty in quantifying financially what is related to climate; myopic view of risks that do not look in the medium to long term but only in the short term. The TCFD listed four topics deemed crucial for climate-related disclosure across all sectors and jurisdictions: *governance, strategies, risk management, metrics,* and *targets* (TCFD, 2017). In developing its recommendations, the Task Force pointed out that there was a problem of comparability of disclosed information as well as a lack of general uniformity in reporting. There was a need to align the different approaches and jurisdictions to achieve a common overall picture. Two items thus seemed particularly relevant: location (i.e., the inclusion of non-financial climate-related information directly within mandatory annual reports) and materiality (i.e., quantitative and qualitative information that is useful to stakeholders and the outside world, but also within organizations themselves). The TCFD, to support its recommendations on climate-related disclosure, listed several principles to make disclosure effective: the information reported should be relevant, concise, complete, clear, understandable, consistent, comparable by sector/industry/portfolio type, reliable, verifiable, objective, timely, and regular over time. The recommendations of the TCFD were intended to provide a solid basis for improving the transparency of climate disclosures and to enable (stakeholders, investors, and governance bodies of each organization) a proper assessment of relevant risks and opportunities.

With reference to this, **New Zealand**, which in the 1990s had not yet introduced specific reporting requirements for environmental concerns, was a pioneering country to consider the TCFD recommendations. A "Terms of Reference" addressed to the Productivity Commission by the Government was issued on April 26, 2017, asking for an investigation into the best way for New Zealand to move toward a reduced net emissions economy. The "2016 Nationally Determined Contribution" was revised on October 31, 2021, to cover the period spanning 2021 through 2030. It established a goal to cut greenhouse gas emissions by 50% below New Zealand's gross 2005 level by 2030. It also specified the type of information that must be disclosed under the Paris Agreement to promote clarity, openness, and understanding. To introduce a framework for climate-related disclosure, the Financial Sector (Climate-related Disclosure and Other Matters) Amendment Act was definitively ratified on October 27, 2021. Although this Act was first enacted on October 27, 2022, "climate reporting entities" will mostly be required to prepare, provide, and update records for their climate statements starting in April 2024. The final Climate Standards, which comprise the Climate-related Disclosures (CS 1), the Adoption of Aotearoa New Zealand Climate Standards (CS 2), and the General Requirements for Climate-related Disclosures (CS 3), have been issued on December 15, 2022. The Climate-related Disclosures Staff Guidance was completed and published on May 1, 2023. It provides guidelines to climate reporting organizations of all sectors on compliance with the three Climate Standards and includes a variety of illustrative examples. The Financial Markets Conduct (Climate-related Disclosures) Amendment Regulations 2023 aim to introduce further record-keeping rules, transitional conditions, and infringement charges also for minor violations under the climate-related disclosure regime. They were issued as a consultation paper and exposure draft by the Ministry of Business, Innovation, and Employment on June 21, 2023. Therefore, tracing very briefly the steps taken, the process undertaken by New Zealand has been constant and subject to continuous updating.

With reference to TCFD, on June 27, 2023, the **Australian Federal Government** issued a (second) consultation paper entitled "Climate-related Financial Disclosure Consultation Paper" to gather comments regarding the development and implementation of requirements for disclosing climate-related financial risks and opportunities. It suggests making reporting mandatory for Australia's largest financial institutions, listed and unlisted companies starting on July 1, 2024. The reporting requirements will eventually apply more progressively to other companies. Therefore, no specific sustainability disclosure requirements are currently in place in Australia. However, there is recognition of the need and urgency to adapt to the development that is occurring globally. Indeed, even Australian institutions recognize the international view that climate change represents a significant risk to the global financial system and includes material risks. One of the tools that can be used to counter and manage financial risks from climate change is disclosure. Moreover, it is recognized that there has been an increase in demand for such information from investors in particular since the TFCD issued its recommendations in 2017. As a result, guidance has been provided by financial authorities calling for transparency in the disclosure of material risks, encouraging the adoption of the TFCD recommendations. Institutions recognize the danger of creating a gap between the Australian situation and the rest of the world due to recent international normative developments. Furthermore, it should also be considered that Australian companies operating in other markets, e.g., in the EU, are required to provide sustainability reporting. To promote the certainty and the comparability of sustainability disclosure, the Australian government has committed to requiring (at least large and financial) companies to provide information concerning financial risks, opportunities, and climate-related plans. Over the past few years, however, there have only been some initial developments in this regard. Some initiatives are mentioned below.

In the 1990s, there were no particular standards or guidelines on environmental accounting and transparency released by the Public Sector Accounting Standards or the Australian Accounting Standards Board (AASB). Similar to the USA's situation, the Statement of Accounting Concepts n. 4 ("Definition and Recognition of the Elements of Financial Statements") and the Statement of Accounting Concepts n. 3 ("Qualitative Characteristics of Financial Information", AASB, 1990) were also implemented concerning reporting on environmental issues, in particular about the issues of liability recognition and disclosure of relevant and reliable information.

In 1998, the Corporations Law was amended to include section 299 (1) (f), then included in the Corporations Act 2001, which required companies (subject to environmental regulation by the Commonwealth, an Australian state or territory) to disclose information about their environmental performance. In addition, the Modern Slavery Act 2018 required companies with at least AUD $ 100 million in annual revenue to disclose information about modern slavery in the supply chain as well. In November 2021, the FRC (Financial Reporting Council), the AASB (Australian Accounting Standards Board), and the AUASB (Auditing and Assurance Standards Board) issued their position on Extended External Reporting (EER), which marked the need for authoritative guidance in this regard. EER highlighted the presence of various forms of non-financial reporting, such as integrated reporting, sustainability reporting, and ESG-related reporting. In December 2021, the Australian Securities and Investments Commission welcomed the creation of the ISSB and promoted the application of the TFCD recommendations for listed companies. This application was also recommended by the AASB as an initial step toward non-financial reporting in March 2022. This represented only part of the general commitment to lead the development of a framework for sustainable finance in Australia. The underpinning principles were: supporting the climate goals of zero net emissions by 2050; improving the quality, comparability, and quantity of information flows; increasing understanding of climate-related disclosure requirements to facilitate enforcement; alignment with internationally adopted reporting practices to reduce the costs companies would have to incur to comply with other frameworks; flexibility of the new framework to adapt to the present system and future developments; and risk proportionality, whereby the information disclosed should be proportional to the extent of the risks perceived by the company. Therefore, the Australian situation is currently very similar to that in the USA and characterized by the predominantly voluntary approach to sustainability reporting, with concrete regulatory proposals still to be developed or finally approved. Industry-led initiatives are also underway in Australia. Therefore, different taxonomies are being developed in different jurisdictions. There is a need for a common basis for jurisdictional requirements and a global baseline standard for sustainability reporting to ensure sufficient informational comparability.

3.7   INDIA AND ASIA BETWEEN THE PROPENSITY FOR HARMONIZATION AND INDIVIDUAL INITIATIVES

Many Asian regulators (Hong Kong, Singapore, Taiwan, and Malaysia) are requiring TCFD-aligned reporting from financial institutions. However, while closely following and examining the implementation of the EU taxonomy framework, individual regulators are favoring the development of local taxonomies. For instance, the Monetary Authority of Singapore (MAS) is examining green taxonomy to simplify the ESG investment process; in Hong Kong, following the publication of the Common Ground Taxonomy (CGT), the inter-agency steering group for green and sustainable finance is working to finalize proposals for the local "green classification" framework.

In particular, climate change represents a major challenge and urgency for both Asia and India, due to the growing threat of extreme weather events such as floods, hurricanes, and temperature increases, among others. India has therefore committed itself to achieving zero emissions by 2070. In recent times, Indian commitments to develop the decarbonization of its economy have been observed, also prompting market players to adopt sustainable business techniques. ESG disclosure can be considered as part of this strategy. With regard to this disclosure, it can be seen that a regulatory evolution has taken place in India. The first step dates back to 2009 when the Indian Ministry of Corporate Affairs issued the NVGs (National Voluntary Guidelines) for the voluntary disclosure of CSR issues.

The next step was the "Business Responsibility Report" (BRR) of 2012, by which the Securities and Exchange Board of India (SEBI) required the top 100 listed companies, identified by market capitalization, to prepare such a report, based on the NVGs guidelines, along with their annual financial statements. The "Companies Act" of 2013 then introduced several requirements for companies on ESG disclosure, in particular by providing for the preparation of a report by the board of directors regarding energy conservation to be included in the annual financial statements, as stipulated in Section 134(m) and further regulated by the "Companies Rules" of 2014. In general, this Act required companies not to focus solely on the interests of their shareholders, but on the welfare of a broader pool of stakeholders. In 2015, the obligation to prepare the BRR was extended to the top 500 listed companies. In the same year, Regulation 34(3), also known as LODR ("Listing Obligation and Disclosure Requirements"), required listed companies to disclose also their risks and opportunities. However, no information was required on the methods used to identify them.

In addition, in 2017, the Indian Banks' Association issued voluntary national guidelines for responsible finance, including a set of general principles for including risk management from ESG factors in the strategies and decision-making processes of financial institutions. For instance, Principle 2 stated that they should also consider an analysis of ESG factors in their investment, loans, and risk management processes to mitigate possible side effects. However, these guidelines, again, do not provide an explicit framework to be used. In 2017, the voluntary adoption of Integrated Reporting was also promoted for the top 500 companies that were already required to prepare the BRR. Finally, the Business Responsibility and Sustainability Report (BRSR) was introduced in May 2021 with the amendment of the aforementioned Regulation 34(2)(f). The BRSR replaced the BRR and its preparation is mandatory from the fiscal year 2022–2023 for the first 1,000 listed companies, involving the disclosure of information related to ESG issues. It is based on the following fundamental principles: companies conduct their business with "integrity", ethics, and transparency; goods and services are provided sustainably and safely; the welfare of employees is ensured and the welfare of workers in the value chain is promoted; the interests of all stakeholders are respected and met; human rights are protected and promoted; and environmental protection and restoration are safeguarded. Furthermore, when companies influence the political and regulatory spheres of the country, they should act responsibly and transparently. Finally, the last two principles state that companies should foster inclusive growth, ensure fair development, and provide value to their customers in a responsible manner. Furthermore, the aim is to make corporate practices more ESG-compliant. To this end, the BRSR requires disclosure of policies and mechanisms that the company can develop to comply with ESG requirements, greater transparency for social and climate-related issues, and a distinction between the disclosure of core indicators and leadership indicators. Finally, it provides for the possibility of voluntary adoption also for companies that are not bound by, but are concerned about, sustainability issues and that wish to use the framework proposed by the SABR. To facilitate the preparation of the report, a guidance document was also published containing three main parts: General Disclosures, Management & Process Disclosures, and Principle Wise Performance Disclosures. Finally, companies that already voluntarily prepare ESG reports based on other international standards (e.g., GRI, TCFD, or CDP) are still required to prepare their report based on the framework provided by the BRSR, although it is expected that it may cross-reference with other documents prepared according to other international standards.

3.8   THE UNIQUENESS OF SOUTH AFRICA'S CASE

Reporting in Africa is limited and plagued by issues that are specific to the area. Firms operating in Africa face accounting challenges such as underdeveloped legal frameworks, difficulties in applying international accounting standards, lack of skill of accounting professionals, and minimal audit infrastructure. Although the well-known economic growth, social and environmental welfare is still worsening in several African regions. Aside from South Africa, sustainability reporting in Africa is a voluntary effort with no governmental standards or regulations. Despite the clear benefits of preparing sustainability reports, African firms are slow to develop them. Nevertheless, the situation in South Africa is special since the Johannesburg Stock Exchange mandates that listed South African corporations submit sustainability information on a "report or explain" basis. In particular, the King Report on Corporate Governance (King III) requires Integrated Reporting (IR). Indeed, the Republic of South Africa has been crucial to the development of IR, especially in regulatory terms, since 1994 with the publication of the King Report I and its subsequent versions: the first version of 1994 was followed by the second of 2002, the third of 2009, and the fourth of 2016, requiring entities listed on the Johannesburg Stock Exchange to compulsorily prepare IR. Wanting to clarify the differences in the meanings of the various concepts involved, and wanting to summarize the relationships between them, we can say that integrated thinking is considered the foundation for the definition of a broader and integrated process of reporting where IR is its final output. IR is not the simple merging of financial and sustainability reporting but becomes an additional annual report. The framework stems from the awareness that knowledge, human resources, and financial resources are now increasingly intertwined in an increasingly unpredictable and ever-changing economic environment. Consequently, IR aims to be more useful and to be able to obtain more consistent investments in the long run. The traditional financial annual report was considered outdated in the context. It was no longer sufficient to adequately meet the informational needs of stakeholders. Among its dysfunctional features was its primarily past-oriented perspective, presenting final results that were of little interest for understanding future trends, and its inability to adequately deal with so-called "intangibles". Other particularly critical points had to be overcome, such as the excessive length and complexity of other types of reports, which were often poorly summarized and therefore not very functional for effectively communicating corporate issues. Many of these same documents, in addition to appearing many times incoherent concerning the entire information system and disconnected from the accounting documentation, proved incapable of adequately presenting the relationships existing between sustainability and the company's basic strategy, making it difficult to compare them. *"These tools were not able to harmonize the various sources of corporate reporting and therefore failed to provide a holistic view of the company's performance"* (Vitolla & Raimo, 2018). IR enters into a confused accountability panorama as a hybrid form capable of bringing together information of different natures to be able to align data with the different needs of capital providers, achieve a better quality of reported information, convey a higher degree of confidence, and pursue a better allocation of resources as well as prompter management of risks (Mio et al., 2022). Precisely, the coexistence of heterogeneous information is the basis of also sustainability reporting. The vision here is not atomistic but holistic, to progressively regain the trust of stakeholders and the community through a transparent and widely comprehensible tool no longer reserved for a few insiders. It is designed to allow communication (and eventually positively change the market perception about) business performance by considering internal organization, taking ESG factors into account, fostering a longer-term perspective, and greater involvement of corporate actors. The IR framework provides a set of general concepts and principles to be followed as a guarantee of reporting quality and not particular analytical performance indicators or rigid standards to be compulsorily adhered to. The IIRC itself explains how such an approach aims to *"strike an appropriate balance between flexibility and prescription that recognizes the wide variation in individual circumstances of different organizations while enabling a sufficient degree of comparability across organizations to meet relevant information needs. It does not prescribe specific key performance indicators, measurement methods, or the disclosure of individual matters, but does include a small number of requirements that are to be applied before an integrated report can be said to be in accordance with the IR Framework"* (IIRC, 2021).

The structure of the IR framework consists of two macro-sections. The first is an introductory part in which the logic and basic approach of Integrated Thinking are set out and in which details are provided on how to use the framework and the basic concepts. This first section is functional to the second, more specific section that provides the specific instructions to be followed for the actual drafting of the IR with indications on the guiding principles and content elements. It follows that different organizations are left to choose which relevant issues to consider and how to report on them, using their judgment in assessing and dealing with different types of matters. This is certainly one of its strengths but also one of its most marked differences from the GRI Standards seen above. The IR framework offers a different way of conceiving and reporting business activity. Indeed, the main innovation brought about by the latest King IV is the "comply and explain" principle, which requires the application of the required criteria and a narrative explanation of how they have been implemented through appropriate practices. The information provided must be detailed and relevant to allow for proper stakeholder evaluation, but leaving some discretion to the management in choosing which information to disclose. Moreover, the Johannesburg Stock Exchange (JSE) had already issued the "Social Responsible Investment Index" in 2004 to promote the disclosure by listed companies of information on their sustainability performance. To date, this index has been replaced by the more advanced FTSE/JSE ESG indices, following the developments that have taken place both at the national level with the various King Reports and at the international level on sustainable initiatives.

3.9 Concluding Remark

The chapter emphasizes some peculiar issues, such as the shift from generic non-financial disclosure to full sustainability reporting. This has been promoted by a growing recognition of the importance of systematizing, institutionalizing, and formalizing the way of providing more transparent information about sustainability issues to external stakeholders, as also demonstrated by the ongoing transition from voluntary to mandatory sustainability reporting. However, to the date still a minority of companies are globally required to provide mandatory sustainability reporting. The lack of global convergence in sustainability regulation, the diverse range of corporate reporting standards currently used around the world, and the proliferation of sustainability frameworks (both issued and adopted) make informational comparison across companies and markets challenging. In this regard, a detailed overview of the usage of sustainability reporting standards and frameworks worldwide emerges from the 2024 edition of the IFAC report "The State of Play: Sustainability Disclosure and Assurance". The report shows that the GRI Standards remain the dominant standard, with an average rate of reporting of 77%. Looking at individual jurisdictions, it emerges that the GRI Standards are utilized by nearly all European companies (100% in Italy and 98% in Spain) while their usage decreases in the USA (75%), South Africa (62%), and the UK (50%). In the USA, the leading reporting standards are the SASB (93%) and the TCFD Recommendations (88%), which have much lower adoption rates in Europe. Furthermore, most companies (87% in 2022) continue to use multiple types of standards and frameworks to report sustainability information. This heterogeneity of adopted sustainability frameworks and standards and the subsequent lack of comparability imply several risks. Indeed, they generate informational costs and barriers that undermine the global capital market, especially in a context where investors increasingly recognize the importance of understanding the sustainability profiles of companies in their investment decision-making processes. Furthermore, the chapter clearly points out that in some countries the disclosure is limited to issues concerning selected areas of sustainability (such as the environment and climate change concerns in the USA), preventing comprehensive sustainability reporting at a time when stakeholders increasingly expect information also on social issues, such as diversity and equity, and governance. Fortunately, alignment seems to be currently underway, both in terms of standards and frameworks and in terms of legislation.

Concerning standards, the chapter has highlighted that the sustainability reporting landscape will be characterized by three main sets of standards: the GRI Standards, ESRS, and ISSB. As mentioned above, the GRI Standards are actually the most adopted and they will be still adopted by organizations that will not be required to provide sustainability reporting. The European Sustainability Reporting Standards (ESRS) have only recently been promulgated, but will play a major role due primarily to European legislation (CSRD, 2022). Finally, the ISSB Standards provide a comprehensive global baseline of sustainability disclosure standards that can be mandated and combined with jurisdiction-specific requirements or requirements aimed at meeting the information needs of broader stakeholder groups beyond investors. Consistent with the approach taken for IFRS Accounting Standards issued by the IASB, it is for jurisdictional authorities to decide whether to mandate use of IFRS Sustainability Disclosure Standards issued by the ISSB. The ISSB's work is backed by the G7, the G20, IOSCO, the Financial Stability Board, African Finance Ministers, and Finance Ministers and Central Bank Governors from over 40 jurisdictions. Following a comprehensive review, the International Organization of Securities Commission (IOSCO) announced its endorsement of the ISSB Standards, encouraging their widespread adoption and sending a strong signal to jurisdictions around the world that they can be confident in implementing the ISSB Standards into their regulatory frameworks. Dozens of jurisdictions around the world are already actively considering their adoption roadmaps and pathways toward mandatory application of ISSB Standards.

Concerning legislation, the chapter has highlighted that the European Union has been the precursor in mandatory sustainability reporting, that was introduced by the Non-Financial Reporting Directive (NFRD, 2014) and recently reviewed by the EU Corporate Sustainability Reporting Directive (CSRD, 2022). The CSRD has not only enlarged the number of companies under the scope of mandatory sustainability reporting, but also mandates these companies to apply the ESRS, a new set of standards developed by the European Commission in collaboration with the European Financial Reporting Advisory Group (EFRAG). At present, the short-term benefits for companies required to report on sustainability according to the "double materiality" perspective are not so evident.

The increase in both the volume and quality of sustainability corporate disclosure aims to improve transparency in environmental, social, and governance issues, to counteract greenwashing, and to strengthen the sustainable orientation of the European market by laying the foundation for sustainability reporting standards on a global level. These effects imply broadening of the "ultimate beneficiaries" of sustainability reporting, including civil society actors, non-governmental organizations, and social partners, who expect companies to be more responsible for their impact on people and the environment. Mandatory sustainability reporting requirements should take more account of the real information needs of stakeholders and the capacity of companies to provide sustainability-related data and information; otherwise, reporting regulation would merely become a checklist to be complied with, imposing high burdens on businesses.

References

AAA (American Accounting Association). (1971). Report of the committee on non-financial measures of effectiveness. *The Accounting Review* (Suppl. to 46), 165–211.

AAA (American Accounting Association). (1975). Report of the committee on accounting for social performance. *The Accounting Review* (Suppl. to 50), 38–69.

AAA (American Accounting Association). (1978). *Report of the committee on the social consequences of accounting information*. AAA.

AASB (Australian Accounting Standards Board). (1990). Qualitative characteristics of financial information. Statement of Accounting Concepts No. 3. ASCPA.

AICPA (American Institute of Certified Public Accountants). (1973). *Objectives of financial statements*. AICPA.

AICPA (American Institute of Certified Public Accountants). (1976). *The measurement of corporate social performance*. AICPA.

CSRD. (2022). Directive 2022/2464/EU of the European Parliament and of the Council of 14 December 2022 amending Regulation (EU) No 537/2014, Directive 2004/109/EC, Directive 2006/43/EC and Directive 2013/34/EU, as regards corporate sustainability reporting.

Deegan, C. (2017). Twenty-five years of social and environmental accounting research within critical perspectives of accounting: Hits, misses and ways forward. *Critical Perspectives on Accounting, 43*, 65–87.

EFRAG & GRI. (2023). *EFRAG-GRI joint statement of interoperability*.

Financial Accounting Standards Board (FASB). (1975). *FASB Statement No. 5, accounting for contingencies*. FASB.

Financial Accounting Standards Board (FASB). (1976). *FASB Interpretation No. 14, reasonable estimation of the amount of a loss*. FASB.

Financial Accounting Standards Board (FASB). (1991a). *Issue No. 89-13, accounting for the cost of asbestos removal*. FASB.

Financial Accounting Standards Board (FASB). (1991b). *Issue No. 90-8, capitalization of costs to treat environmental contaminations*. FASB.

Financial Accounting Standards Board (FASB). (1995). *Issue No. 93-5, accounting for environmental liabilities*. FASB.

Gamble, G. O., Hsu, K., Jackson, C., & Tollerson, C. D. (1996). Environmental disclosures in annual reports: An international perspective. *The International Journal of Accounting, 31*(3), 293–331.

Grewal, J., Hauptmann, C., & Serafeim, G. (2021). Material sustainability information and stock price informativeness. *Journal of Business Ethics, 171*, 513–544.

GRI. (2023). *Universal Standards. Setting a new global benchmark for sustainability reporting. They are in effect for reporting from 1 January 2023*. https://www.globalreporting.org/standards/standards-development/universal-standards/

Hahn, R., & Kühnen, M. (2013). Determinants of sustainability reporting: A review of results, trends, theory, and opportunities in an expanding field of research. *Journal of Cleaner Production, 59*, 5–21.

Haller, A., Link, M., & Groß, T. (2017). The term 'non-financial information'— A semantic analysis of a key feature of current and future corporate reporting. *Accounting in Europe, 14*(3), 407–429.

Harper Ho, V. (2020). Non-financial reporting & corporate governance: Explaining American divergence & its implications for disclosure reform. *Accounting, Economics, and Law: A Convivium, 10*(2), 20180043.

International Integrated Reporting Council (IIRC). (2021). *International <IR> framework*. https://integratedreporting.ifrs.org/wp-content/uploads/2021/01/InternationalIntegratedReportingFramework.pdf

Mathews, M. R. (1997). Twenty-five years of social and environmental accounting research: Is there a silver jubilee to celebrate? *Accounting, Auditing and Accountability Journal, 10*(4), 481–531.

Millstone, S., & Watts, F. B. (1992). Effect of the green movement on business in the 1990's. *The Greening of American Business*, 1–31.

Mio, C., Agostini, M., & Panfilo, S. (2022). Bank risk appetite communication and risk taking: The key role of integrated reports. *Risk Analysis, 42*(3), 634–652.

NAA (National Association of Accountants). (1974, February). Report to the committee on accounting for corporate social performance. *Management Accounting*, 39–41.

NFRD. (2014). Directive 2014/95/EU of the European Parliament and of the Council, 22 October 2014, amending Directive 2013/34/EU as regards disclosure of non-financial and diversity information by certain large undertakings and groups.

Pizzi, S., Principale, S., & de Nuccio, E. (2022). Material sustainability information and reporting standards. Exploring the differences between GRI and SASB. *Meditari Accountancy Research, 31*(6), 1654–1674.

Pizzi, S., Caputo, F., & De Nuccio, E. (2024). Do sustainability reporting standards affect analysts' forecast accuracy? *Sustainability Accounting, Management and Policy Journal, ahead-of-print*. https://doi.org/10.1108/SAMPJ-04-2023-0227

SEC (Securities and Exchange Commission). (2024). The Enhancement and Standardization of Climate-Related Disclosures for Investors, Release Nos. 33-11275; 34-99678; File No. S7-10-22, RIN 3235-AM87 (Washington D.C., March 6, 2024).

Task Force on Climate-related Financial Disclosures (TCFD). (2017). *Recommendations of the task force on climate-related financial disclosures*. https://www.fsb-tcfd.org/recommendations/

United Nations. (1992). *Environmental accounting, current issues, abstracts and bibliography*. United Nations.

Vitolla, F., & Raimo, N. (2018). Adoption of integrated reporting: Reasons and benefits—A case study analysis. *International Journal of Business and Management, 13*(12), 244–250.

Wiseman, J. (1982). An evaluation of environmental disclosures made in corporate annual reports. *Accounting, Organizations, and Society, 7*(1), 53–63.

Non-financial Disclosure and Sustainability Reporting: A Systematic Literature Review

Abstract The evolution and corporate awareness of non-financial disclosure and sustainability reporting have notably increased in recent years, as emphasized in the academic literature. In order to examine how the academic literature has dealt with non-financial disclosure and sustainability reporting, a systematic literature review has been carried out by searching relevant publications in the Scopus database and following a consecutive-steps approach of analysis. The results of the literature review subsequently conducted are divided into paragraphs, indicating the issues most considered. These concern: the designation and media of sustainability disclosure; the need to disclose both significant and reliable information about the materiality assessment process carried out upstream by the organization; the function and motivation of sustainability reporting, recalling the different theoretical approaches that support them; the relationship between disclosure and corporate performance; the commitment of providing both quantitative and qualitative information through sustainability reports, highlighting also the increasing relevance of assurance.

Keywords Systematic Literature Review · Sustainability reporting · Assurance

© The Author(s), under exclusive license to Springer Nature Switzerland AG 2024
C. Mio et al., *Sustainability Reporting*, Palgrave Studies in Impact Finance, https://doi.org/10.1007/978-3-031-58449-7_4

4.1   ANALYSIS: LITERATURE EVOLUTION

The evolution and corporate awareness of non-financial disclosure and sustainability reporting have notably increased in recent years (KPMG, 2017, 2022), as facilitated and emphasized in the academic literature. Specifically, a consistent number of studies have focused on the evolving EU normative context following two main directions of analysis. The first stream of literature examines corporate adjustments implemented to comply with evolving national regulations (Dumitru et al., 2017; La Torre et al., 2018; Venturelli et al., 2017). The second analyses the transposition of the EU directive into national laws, focusing on the degree of compliance and the achievement of formal harmonization across EU Member States (Aureli et al., 2019). There are still various issues concerning EU directives that have not yet been explored (Biondi et al., 2020). Corporate reporting appears to be characterized by a varying degree of disclosure in terms of both the volume and quality of NFI (Brammer & Pavelin, 2008; Rezaee & Tuo, 2019), with a consequent decrease in stakeholders' trust in reported information (Chaidali & Jones, 2017; Dumay et al., 2019). The literature stresses the importance of the quality and transparency of NFI (Aureli et al., 2019; La Torre et al., 2018). This purpose of increasing information quality and improving stakeholders' trust has been pursued through new normative provisions, even though some literature shows that regulation is not always an effective way to change behaviors and practices (Bebbington et al., 2012; Luque-Vilchez & Larrinaga, 2016). There is a coexistence of mandatory and voluntary aspects in terms of both normative approaches and reporting means because of the peculiar separation and/or integration of coercive and voluntary disclosure (Bold, 2017; Delbard, 2008; Doshi et al., 2013; Martin-Sardesai & Guthrie, 2019) that is now mandatory in nature (stemming from the latest EU provision) but still voluntary in terms of being driven by stakeholders' expectations (O'Donovan, 2002) and firm characteristics (Clarkson et al., 2008). Coherence is needed to eliminate incongruences of sustainability reporting from both theoretical and practical perspectives; otherwise, legal requirements become impractical and illegitimate (Bebbington et al., 2012). This issue concerns a wide community of stakeholders in a decisive period of transition that is characterized by the coexistence of mandatory and voluntary normative approaches concerning sustainability reporting. Non-financial reporting has been the subject of multiple studies; several authors have analyzed it from different points of view. The purpose of this chapter is to systematically analyze the contributions in the literature concerning non-financial disclosure and sustainability reporting.

4.2   Research Methodology: Systematic Literature Review

To examine how the academic literature has dealt with non-financial disclosure and sustainability reporting and to understand the effects of the latest normative provision a systematic literature review has been carried out by searching relevant publications in the Scopus database and following a consecutive-steps approach of analysis (Littell et al., 2008; Massaro et al., 2016; Sivarajah et al., 2017). Papers most related to the topic were identified through the use of appropriate keywords to search the database. Specifically, the following string was used:

TITLE-ABS-KEY (sustainab* OR non-financial AND disclos* AND report* OR account*)

The use of the keywords *sustainab**, *non-financial*, and *disclos** aim to identify works that relate to non-financial disclosure and/or about sustainability issues. The explicit reference to the terms *report** and *account** aims to circumscribe the results of the literature search to the field of accounting, focusing on the means (corporate reports and/or accounting tools in general) by which such information is or should be disclosed. The search string identified 3798 papers in the Scopus database. However, it was necessary to delimit the scope of the search by introducing some objective parameters, thus going on to identify the results most inherent to the topic. First of all, only those papers were selected (through the appropriate setting provided by the Scopus database *"Document type: Article"*) that were definitively published (*"Publication stage: Final"*) within scientific journals (*"Source type: Journal"*), thus reducing the results of the bibliographic search to 2803. To enable full understanding and comparison, all papers written in a language other than English (*"Language: English"*) were excluded, having confirmed that most of the articles (2713) were written in English.

The next step was to select papers from the perspective of their subject area. Using the *"subject area"* function, it was possible to limit the search to papers belonging to the subject areas *"Business, Management and*

Accounting". These are the most relevant to the topic covered in this book and include the largest number of contributions (1654 papers).

Finally, to carry out a relevant analysis based on the papers deemed most significant in the literature, it was decided to keep only those works published in the highest quality academic journals according to the ABS 2021 list prepared by the "Chartered Association of Business Schools": for the present analysis, journals in the *"account"* field falling in bands 4*, 4, and 3 were considered. The results thus obtained are 135. The analysis was then based on reading the title and abstract of the 135 papers thus identified, to ascertain which papers were most relevant to the topic on which this work focuses.

4.3   Systematization and Categorization: Literature Streams and Results

The analysis of the sampled papers leads to their grouping according to different perspectives. In this way, several subtopics are identified to better examine and understand the impact of the normative evolution on non-financial disclosure in corporate reports, the theoretical framework(s) that underpin this relationship, and provide future avenues of research. The literature review subsequently conducted is divided into paragraphs, indicating the macro-topics in which it is possible to fit in various contributions, although many papers deal with different topics. The authors focus on multiple aspects of non-financial reporting, however, the issues most considered concern: the naming and location of sustainability disclosure, the reporting of which has long been considered exclusively voluntary and flexible; the need to disclose both significant and reliable information about the materiality assessment process carried out upstream by the organization; the function and motivation of sustainability reporting, recalling the different theoretical approaches that support them; the relationship between disclosure and corporate performance; and the importance of providing both quantitative and qualitative information through sustainability reports.

Designation and Media for Sustainability Disclosure

The last few decades have witnessed a substantial increase in the use by larger companies of social and environmental reporting. Stolowy and Paugam (2018) verified that, for listed companies in the UK and EU, the percentage of those publishing a sustainability report increased significantly, from 6% in 2006 to 77% in 2015. However, before recent European regulatory measures, the term "sustainability" was rarely and vaguely used in corporate reports (Rimmel & Jonäll, 2013). In addition, a clear difficulty emerged in applying a common designation to disclosures dealing with non-financial and sustainability issues. Indeed, initially the term "social and environmental reporting" was employed, but in recent years it has become infrequently used (Farneti & Guthrie, 2009), gradually giving way to other designations such as corporate social responsibility reports, "non-financial reporting," or "sustainability reporting" (Adams & Larrinaga-González, 2007; Stolowy & Paugam, 2018). Indeed, most works between 1973 and 2013 recalled corporate social responsibility (CSR) to name reports about NFI (Erkens et al., 2015). In addition, the different naming was also intended to signal the different focus of documents prepared by companies to communicate primarily socio-environmental issues, to highlight the distinction from the actual corporate annual report, and to emphasize corporate communication provided without precise regulatory guidance (Thorne et al, 2014). However, the term "sustainability reporting," where adopted, seemed to be understood in very general terms, with no clear convergence of meaning and/or common framework. Indeed, even studies based on interviews and empirical analysis showed that managers did not have an unambiguous definition of sustainability, even with the understanding that it incorporated not only environmental issues but also social, ethical, and political ones (Farneti & Guthrie, 2009). Therefore, "triple-bottom-line reports" and "social accounting reports" could be considered corporate sustainability reports (Leong & Hazelton, 2019).

In addition to the designation of the report, the location of the information being disclosed, thus the communicative means used to make that disclosure, has also been a subject of debate in the literature. One strand of research has long argued that the annual report is an excellent means of communication even for NFI (Farneti & Guthrie, 2009). Another strand of research has long argued that CSR disclosure makes use of stand-alone documents that can take many forms, but usually take the form of reports, whether voluntary or mandatory, that are nonetheless annual (Thorne et al., 2014). Indeed, the two strands are not in opposition because the literature shows that there has been a real evolution of such reporting pertaining to NFI, starting from the increased information disclosed in annual reports to the preparation of independent and specific reports, the number of which is constantly growing (Buhr, 2002; Cho et al., 2015b; Michelon et al., 2015; Milne & Gray, 2007). In addition to the report, a key role for non-financial disclosure is played by the website: typically, companies, that publish the report, also reserve a special section of their website for the communication of information about sustainability, and in most cases, the content of this section turns out to be identical to that proposed in corporate reporting (Rimmel & Jonäll, 2013). In addition, sometimes special communicative emphasis and appropriate placement (including on the corporate website) is given to issues considered by companies as particularly relevant (Adler et al., 2018). Sometimes again, due to the flexibility and voluntariness of sustainability disclosure, different placements are provided for different types of information and subject areas (Adler et al., 2017; Frost et al., 2005; Guthrie et al., 2008).

The Materiality of Disclosed Sustainability Information

In addition to the designation given to different forms of reporting and the location deemed preferable for the disclosure of NFI, the literature has long emphasized the need to establish a process aimed at determining the materiality of disclosed sustainability issues to increase the transparency of information, the quality of reports, and the trust of stakeholders (Beske et al., 2020; Dilling & Harris, 2018; Eccles et al., 2012; Fasan & Mio, 2017; Steenkamp, 2018). Suggestions provided by the various voluntary standards and frameworks published over the years have proven useful and appreciated in the literature (see this chapter for a focus on materiality). In general terms, materiality implies that all information that significantly influences, and is preponderantly relied upon by, stakeholder assessments is provided (GRI, 2013). Given the multitude of information that could be considered pertaining to sustainability, assessing materiality becomes essential for recognizing and reporting corporate distinctiveness. Doing so allows the amount of redundant information to be reduced (Accountability, 2013; IIRC, 2013). However, wanting to go into more detail, the term "materiality" itself is not easy to define, as highlighted and explored also in this chapter. It was initially devised concerning traditional financial reporting, which aims to provide a fair and truthful representation of corporate information to support the economic decisions of the users of financial statements, primarily believed to be investors. However, even at the non-financial level, the literature has highlighted the relevance of the materiality principle in meeting the informational demands of stakeholders regarding economic, social, and environmental impacts (Fasan & Mio, 2017; Lai et al., 2017; Puroila & Mäkelä, 2019; Steenkamp, 2018). Practitioners have long been called upon to carry out an evaluation process aimed at determining what information (not pertaining only to economic issues) is to be considered most meaningful, with what indicators it can best be expressed, what range of stakeholders (certainly broader than just investors) can be satisfied (O'Connor & Spangenberg, 2008; Owen et al., 2001). Similar to the financial perspective, even in the context of sustainability, the assessment of materiality therefore requires subjectivity and necessitates professional judgment. It therefore takes on the "appearance" of an opinion rather than a technical and mechanical process (Gray & Manson, 2008), requiring the assessment of several items about what, who, how, and why (Brown & Dillard, 2014). A key element, also capable of bringing greater objectivity to this evaluation process, is stakeholder engagement, which should require dialogic accounting (Puroila & Mäkelä, 2019). The literature shows that different frameworks (e.g., GRI, IIRC, SASB) voluntarily adopted by companies have led not only to different concepts of materiality but also to different corporate stakeholder engagements. To increase perceived trustworthiness concerning disclosed issues, stakeholders need to know the process that led the organization to recognize such issues as material (Cerbone & Maroun, 2020; Farooq & De Villiers, 2019; Ruiz-Lozano et al., 2022). Transparency about the materiality assessment process is relevant to increase its credibility (Steenkamp, 2018). Despite this relevance, the disclosure of information about the materiality process is not yet adequately developed, being still at an early stage. In general, the identification of material information to be disclosed is based on the information available to the organization, stakeholder expectations, corporate and industry risks and opportunities, and the various reports disseminated by the company (Ruiz-Lozano et al., 2022). Even where they are effectively identified, information regarding material issues is not always then conveyed due to some factors highlighted in the literature (Farooq & De Villiers, 2019). A few are mentioned as examples. First, according to legitimacy theory, managers show reluctance to disseminate information about negative issues even if they are material. This difficulty factor becomes even more relevant if competitors do not make a voluntary disclosure of sustainability information, so the disclosure of non-financial, sensitive, and negative information is perceived as a source of competitive disadvantage. Second, obtaining certain data can sometimes prove too costly, especially if that cost is not reflected in high stakeholder interest. Third, sometimes the information to be disclosed requires data that are not available to the company, which would have to request them from third parties, who in turn may not know or be involved in the practice of sustainability reporting. With reference to this, corporate groups display difficulties associated with data retrieval and responsibility for preparing the sustainability report. The materiality process can thus become complex and sometimes lack the necessary guarantees to support the reliability of the information disclosed. Finally, the influence exerted by the media may lead the company to favor the disclosure of information on particular issues at the expense of others, regardless of the materiality assessment previously made. Therefore, there emerges the need to disclose not only material information on sustainability matters but also reliable information on the materiality assessment process carried out upstream by the company. It is also crucial in light of the latest European provisions analyzed in the previous chapter.

Function and Motivation of Sustainability Reporting

Most of the literature reviewed deals specifically with the rationale behind sustainability reporting, dwelling on its function in a context in which sustainability is developing and increasingly becoming a crucial aspect. The literature has long stressed the importance of adopting a proactive corporate sustainability strategy (Bouten & Everaert, 2015). However, today there are a multitude of motivations that can drive a company to prepare (even voluntarily) reporting on sustainability issues.

First of all, the current historical period has made evident the situation of climate change, declining biodiversity, degradation of ecosystems, the effects of pollution, and many other critical socio-environmental factors. The threats, especially environmental threats arising from ongoing change, pose a real and now visible risk to the planet. Human beings are both participants and observers of this environmental evolution and must seek to change course by taking into account the current and future impacts of their activities (Jones, 2010). This represents a primary role attributed to sustainability reporting by which companies should adequately account to their stakeholders for their efforts, given also the growing demand for information in this regard (Milne et al., 2009).

Considering, moreover, that it is a society that legitimizes industrial activity, the latter must act to protect the former. Thus, it can be grasped that companies hold a moral responsibility to society, from which they must obtain consent (legitimacy) to operate. Here, then, is the need for a reporting system of sustainability issues that enables the company to perform its *"stewardship"* function by disclosing information regarding the impacts of the company's activities through a holistic accounting system (Jones, 2010). The motivations that may lead a company to prepare (even voluntarily) reporting on sustainability issues are traced in the literature to different theoretical approaches, some of which are recalled below.

The Theory of Voluntary Disclosure

Voluntary disclosure of NFI has played a key role in recent years, as well as pointed out in the literature. According to the voluntary disclosure theory, companies voluntarily decide to disclose information about specific issues to compensate for the information asymmetry between managers (principals) and investors (agents). Indeed, some authors have long emphasized the importance of including information about environmental issues in financial statements, primarily through their numerical-monetary representation. Such voluntary disclosure can also take place through other means than financial statements, such as independent reports and websites (Cho & Patten, 2013). In doing so, companies can distinguish themselves (by the degree of commitment, information transparency, and more satisfactory impacts) and try to boast a "superior market position". This also invokes the *Signaling theory* as sustainability reporting and disclosure of NFI signal corporate values to the public and ensure that corporate stakeholders are fully aware of them. This has led the literature to analyze the relationship between performance and sustainability disclosure. Companies that implement a proactive sustainability strategy are expected to perform better in this regard and will tend to disclose more information for reporting value pursued and activity accomplished. Conversely, lower levels of sustainability performance will imply less information disclosed in an attempt to hide, in whole or in part, these corporate (bad) achievements (Clarkson et al., 2008). In general, the theories primarily adopted to explain and motivate sustainability reporting are socio-political theories, including legitimacy theory and stakeholder theory (Patten, 2002).

The Theory of Legitimacy

Based on legitimacy theory, companies capable of conveying their image with sustainability awareness through appropriate voluntary disclosure can reduce both the likelihood of incurring social-political costs and external pressures from stakeholders (Adams, 2002; Tate et al., 2010). Indeed, legitimacy involves the general assumption that an entity conducts its business desirably and appropriately within the socio-normative context characterized by certain values, beliefs, and definitions (Hrasky, 2012; Suchman, 1995). A company's development depends on its ability to create a desirable outcome and distribute economic, social, and political benefits to its stakeholders (Lindblom, 1994). As a result, firms that are perceived as legitimate are deemed trustworthy and are supported by society; in contrast, firms lacking legitimacy are likely to struggle to obtain the resources they need to survive (Meyer & Rowan, 1991). Therefore, legitimacy is not a right due to organizations from their inception, being instead conferred by society when the organization is perceived as aligned with the shared values of society (Lindblom, 1994). Therefore, on the subject of sustainability disclosure, it is evident how legitimacy theory explains how, when, and why certain information is (even voluntarily) disclosed by the company. However, such legitimacy, once conferred, does not represent a permanently acquired right and can be revoked when the organization breaks the abstractly stipulated social contract (Deegan, 2002). There is a dynamism inherent in the concept of legitimacy that can change over time (Makela & Nasi, 2010). In the current context in which sustainability is a central matter for corporate legitimacy, companies that do not commit to its protection risk "breaking" the implicit contract with society and thus being considered devoid of social value (Deegan et al., 2002; Hrasky, 2012). This corporate quest for legitimacy can result in both positive and negative effects, when, for example, companies prepare sustainability reports aimed at pursuing their interests and masking negative socio-environmental performance (Milne & Gray, 2013). To gain legitimacy, a company can adopt mainly two strategies, i.e., *substantive management* or *symbolic management* (Hopwood, 2009). The first implies that companies, also in strategic response to external stimuli, attempt to gain legitimacy through actual variation in their business practices and policies. The second, on the other hand, concerns the corporate attempt to manipulate stakeholders' perceptions through the disclosure of information designed to demonstrate a corporate commitment that is not effective and has no concrete feedback. In this second case, corporate disclosure serves to create or strengthen a kind of "veil" around the company, which in turn generates a corporate image that is different from reality (Hopwood, 2009). Beyond the strategy adopted (i.e., *substantive management* or *symbolic management*), the legitimacy sought by a company can be pragmatic, moral, or cognitive. The literature has recently shown a link between the strategy adopted, the type of legitimacy sought, and the activity carried out by the firm. In particular, companies operating in more intensive and impactful business sectors promote a strategy aimed at obtaining moral legitimacy through substantive disclosures (Hrasky, 2012). Thus, for example, companies operating in more polluting industries consider environmental disclosures as a means of gaining legitimacy (Barbu et al., 2022). If there is a divergence between corporate and community values, the company will risk losing its legitimacy. Voluntary disclosures can contribute to the strengthening of corporate image (Thorne et al., 2014), demonstrating a newfound realignment of values desirable to stakeholders even in the aftermath of a corporate scandal (Bellucci et al., 2021; Blacconiere & Patten, 1994). Indeed, scandals have the potential to erode corporate legitimacy because they generate the perception of corporate irresponsibility and/or dishonesty, resulting in reputational damage. The so-called legitimacy gap represents the difference between corporate behavior and public expectations (Lin-Hi & Muller, 2013). To close it and reduce the gap created, companies need to restore congruence between their actions and the information disclosed about them (Buhr, 1998; Cho et al., 2012; Lindblom, 1994). Therefore, to create, improve, and restore corporate legitimacy, sustainability reporting can be employed to correct misunderstandings that stakeholders have about the company's business, deflect their expectations (if the company considers them unrealizable), show improvements achieved, and divert attention from particular negative situations to other issues. The strategic use of sustainability disclosures has been considered by numerous studies and, according to legitimacy theory, implies that companies convey information about the management of issues that are particularly relevant to stakeholders (Magness, 2006). Indeed, if a company promotes a proactive and theoretically desirable sustainability strategy, but has a bad reputation because it has been involved in a relevant scandal, its legitimacy is eroded and stakeholders' skepticism is relevant. In this case, the concept of "hypocrisy", associated with "corporate legitimacy", becomes central. Sustainability reporting may restore corporate legitimacy, but it must be appropriate.

The company could deny the occurrence of the scandal, keep silent about it (omit it), or report what happened. This difference in communication may also depend on the situation in which the organization finds itself. For example, if the incident is not the subject of particular attention after a while, the company may decide to keep quiet so as not to attract attention from stakeholders. Conversely, if the incident is particularly monitored by stakeholders, the company might decide to respond to requests made to it and be transparent to seek moral legitimacy (Bellucci et al., 2021). Thus, the diversity of corporate response and behavior depends on a variety of factors also related to the type of scandal that involved the company, the features of the corporate stakeholder group, and the attention shown to the incident by stakeholders and third parties (e.g., the media, social movements, NGOs).

Therefore, the company may pursue a truly deliberate and conscious disclosure strategy, aiming to influence stakeholder perceptions and impressions to protect its reputation through voluntary sustainability reporting (Bowen et al., 2005; Merkl-Davies & Brennan, 2007). The advantageous choice of only favorable content to disclose represents opportunistic managerial behavior aimed at manipulating stakeholder impressions (Miles & Ringham, 2020). Corporate management can use voluntary sustainability disclosures to influence stakeholders' impressions and perceptions about the role the company plays, its responsibility, and how it compensates for it. Therefore, according to an authoritative part of the literature, mandatory sustainability disclosures become relevant because the voluntariness of the information disseminated leads to potentially misleading representations, both for investors and for society as a whole community (Cho & Patten, 2013).

Stakeholder Theory
Another theory that can be applied to explain the reasons for the voluntary preparation of sustainability reports is stakeholder theory, which is closely related to the legitimacy theory discussed in the previous section. The focus is on the consideration of corporate stakeholders whose interests are essential to an organization's survival and prosperity (Adler et al., 2017). Corporate stakeholders can be identified as a unique social group that is influenced and, in turn, influences the organization through different relationships (Crane & Ruebottom, 2011). Underlying this theoretical framework is the assumption that the company is

accountable to its shareholders, but it is also required to balance the interests of all its stakeholders and not just shareholders (Freeman, 1984). Initially, the concept of corporate stakeholder included only creditors and shareholders, i.e., purely financial stakeholders (Gray et al., 1995). Subsequently, the range of corporate stakeholders has gradually broadened to include many others, such as insurers, suppliers, consumer associations, and regulatory authorities. In particular, recent literature has pointed out that the concept of "stakeholder" includes the environment and future generations who, having no voice, are effectively ignored (Bebbington & Larrinaga, 2014; Bouten et al., 2011). In this way, there is no single definition of "stakeholder" because this figure can take on different contours that require deep reflection (Miles, 2017). Sustainability reports play a major role by consisting of a recognized tool that greatly facilitates dialogue between the company and its stakeholders by aiming to respond to the information demands of the latter (Gray et al., 1995; Roberts, 1992; Rodrigue et al., 2013). Stakeholder interests that play a key role in the company will be the first to be satisfied through appropriate disclosure. Stakeholders perceived as "relevant" or "most influential" will be given greater consideration, while less involvement will be devoted to stakeholders perceived as less influential (Haque et al., 2016). Specifically, the literature investigating the reasons for voluntary disclosure (or non-disclosure) of sustainability issues has highlighted the primary importance of companies' willingness to respond to external pressures stemming from stakeholders expressing needs and interests inherent in sustainability practices and policies, sometimes even putting the actual effectiveness and practical efficacy of these practices on the back burner. Even the investigated companies that do not yet prepare sustainability reports claim that they might start preparing them the moment they perceive such external demand and pressure arising from stakeholders or specific regulations (Thorne et al., 2014). Thus, the reasons behind the spread of voluntary sustainability reporting are largely attributable to stakeholders, broadly understood and irrespective of whether there is a formal or regulated relationship between the company and its stakeholders, which can also be the government, communities, environmental groups, or employees (Clarkson et al., 2011; Farneti & Guthrie, 2009). Voluntary sustainability reporting is intended to allow easier access to information considered necessary by stakeholders, as well as to improve the company's accountability and propensity for disclosure. The information needed by stakeholders and provided by the company has been evaluated as material. Indeed, a proper application of the materiality principle (Gray, 2000) is required: stakeholders should be involved in determining what information to disclose and how to broaden the range of users who can have access to it to derive informational benefit. This involvement signals how companies include stakeholders within their decision-making processes, share information, create effective dialogue, and promote mutual accountability. Indeed, stakeholders may also ask the company to bring attention to issues that could have impacts (positive or negative) and, consequently, affect managers' choices (Windsor, 2017). Stakeholder engagement and dialogue are central to sustainability reporting, which aims to become a dialogic process based on co-responsibility relationships between organization and stakeholders (Gray et al., 1997). Dialogic Accounting (DA) represents a form of critical accounting that tries to give adequate representation to the interests of corporate stakeholders, usually not considered in traditional reports (Bellucci et al., 2019; Brown & Dillard, 2015). It aims to gather and make coexistent the different points of view and expectations of various stakeholders (Brown & Dillard, 2013; Gray, 2002). Voluntary sustainability reporting allows stakeholder assessments to be combined in a kind of forum in which freedom and diversity of opinion, social learning, and contestability can be ensured (Brown, 2009; Brown & Dillard, 2014; Dillard & Roslender, 2011). The degree of stakeholder participation and involvement varies from company to company. However, voluntary sustainability reporting rarely highlights the actual difficulties encountered in engaging stakeholders and the concrete ways adopted to facilitate their interactions, revealing little about the actual opinions and involvement of stakeholders in the "debate" promoted by the company on sustainability. Instead, what emerges is the role attributable to new communication channels and in particular social media in developing DA, allowing managers to obtain new data and information from unconventional sources (Manetti & Bellucci, 2016). These forms of communication inherent to sustainability generate useful tools for the company to gain legitimacy. This goal on the one hand seems to push the company to respond more accurately and appropriately to stakeholder information requests by providing relevant information (both positive and negative); on the other hand, it emerges that companies are more likely to omit disclosing information about negative issues that may erode their legitimacy, generating sustainability reports comparable to utopian depictions of their organizations. DA should foster a real and effective debate concerning also the difficulties encountered in the stakeholder engagement process, providing a more reliable and trustworthy image of the organization to reinforce and increase confidence in even the most positive NFI disclosed about the company (Bellucci et al., 2019).

Institutional Theory
Sustainability reporting may represent a corporate tool and the result of precise social pressures (Bebbington et al., 2009; Bouten & Everaert, 2015). These pressures can be of three main types (coercive, regulatory, and mimetic) or they can be a combination of these three types, leading to the achievement of isomorphism (DiMaggio & Powell, 1983).

The three types of pressure have different sources and stem from law, culture, and managerial morality, respectively. They may lead to emulating other organizations that have already implemented sustainability reporting (Barone et al., 2013). Making specific reference to the European context, which is characterized by increasing regulatory expectations and provisions regarding sustainability, the literature finds the presence of coercive, normative, and mimetic isomorphism in firms that aim to imitate other organizations that are already socially legitimized, showing how coercive isomorphism can be reinforced by normative or mimetic one (Barbu et al., 2022). In addition to isomorphism, which is used to indicate homogeneity of corporate response and action, some studies delve into isopraxism, which aims to highlight the differences that can arise when the transmission and spatial shift of a particular corporate idea or practice generate different interpretations or implementation (Adams et al., 2016). Isomorphism and isopraxism, based on institutional theory, are essential to understanding how approaches to sustainability reporting may converge or differ across organizations (Adams et al., 2016).

The differences that there may be in the responses implemented by different organizations can be explained by resistance to institutional conformity. These resistances can be traced to insensitivity, which may be toward the symbolic (such as norms and beliefs) or material (such as systems, routines, artifacts) means of the prevailing logic. Specifically, if organizations do not demonstrate sensitivity to the institutional vectors (symbolic or material) of the prevailing logic, they develop an immunity (symbolic or material) to such prevailing logic. If, on the other hand, they remain attentive and sensitive to such vectors of the prevailing logic, resistance (symbolic or material) to the emerging logic materializes (Lepoutre & Valente, 2012). However, merely symbolic immunity is not sufficient to ensure that organizations follow different and emergent logics, as there may also be material obstacles, such as a stereotypical strategy, routines, and conventional relational systems. These concur to reinforce adherence to the prevailing logic. Applying it to the topic of this book, significant issues that may contribute to hindering the adoption of sustainability reporting may be, for example, limited availability of human and economic resources or the absence of adequate dissemination capacity (Bouten & Everaert, 2015). More generally, using institutional theory, some authors (Leong & Hazelton, 2019) point out that firms do not possess complete freedom of choice for their business decisions. According to this theoretical framework, "the institution" means the set of items that can influence an organization's behavior and decisions. It can include regulatory sources, market pressures, and executive forces (e.g., the intention to avoid a scandal). According to institutional theory, organizations are subject to the influence of a variety of institutions that exert even conflicting pressures (such as concern for the environment and the executive need to maximize profits) to favor specific practices, actions, and approaches (De Villiers & Alexander, 2014; DiMaggio & Powell, 1983). Within this framework, sustainability reporting can help stakeholders participate in "negotiations" and set the organization's agenda, showing real potential to bring about organizational change. Mandatory sustainability reporting appears to be more likely to direct change, provided certain conditions are met, e.g., the indicators used are appropriate for the users of the information, the information provided is aggregated to an appropriate level, the data provided is still comparable with external benchmarks or those of other companies, there is popular and political support that is perceived to be sufficient (Leong & Hazelton, 2019).

In addition, social actors, such as organizations as well, are not only meant to be subjected to institutional pressures, but, if properly educated and equipped with particular skills, can help create, maintain, or destroy an institution (Farooq & De Villiers, 2019; Lounsbury, 2008). In this process, sustainability reporting managers (SRMs) play a key role that needs to be strengthened through appropriate evolutionary growth that can be articulated in four successive phases (Farooq & De Villiers, 2019). The first phase involves training managers, who are still inexperienced, and supporting internal sustainability reporting. The latter is carried out through a discursive strategy to convince that the new practice does not differ much from other responsibilities already received, thus adopting a mimetic approach. The second phase focuses on the transition from a centralized to a decentralized approach to preparing the sustainability report, thus relying on a more engaging process that also involves lower-level managers. The third phase considers the preparation of the sustainability report using a "less is more" approach, including more concise, focused, and materiality-based content. The materiality assessment process becomes primary, formal, and sophisticated, involving the collection of information from numerous internal and external sources. Finally, in the last phase, SRMs are committed to making this reporting practice no longer annual, but an ongoing activity integrated into planning decision-making and involving objectives, compensation, and performance evaluation, through the introduction of appropriate KPIs.

The Relationship Between Sustainability Disclosure and Corporate Performance

The literature shows that a gap continues to persist between the sustainability "talk" perpetrated by organizations and the practical action taken in this regard (Malsch, 2013; Milne & Gray, 2013). This issue is closely related to the main theories underlying sustainability reporting outlined in the previous section. In particular, the relationship between form (the information disclosed) and substance (the practices implemented) can be explained and traced back to the voluntary disclosure theory and socio-political theories mentioned above.

On the one hand, according to the former (i.e., voluntary disclosure theory), the disclosure-sustainability performance relationship should be positive: through the disclosed information, organizations aim to show their achievements by "hard" (i.e., verifiable) means, being superior to those of other entities that, having performed poorly and inferiorly, will not be able to imitate the indicators manifested by the former and will prefer to remain silent about it, hoping to be labeled as "average companies" (Clarkson et al., 2008; Dye, 1985).

On the other hand, the second stream of theories (i.e., socio-political theories) hypothesizes the opposite (i.e., negative) relationship between corporate disclosure and sustainability performance because organizations with poor sustainability performance will tend to disclose more information through "soft" (i.e., less verifiable) means in response to stakeholder pressures to mitigate the risk of loss of legitimacy, to change external perceptions, rather than disclosing their actual (poor) performance (Patten, 2002). The literature has long highlighted the difficulty of finding correspondence between information disclosure and actual sustainability performance. Indeed, some studies have shown that more polluting companies tend to disclose more environmental information, in line with socio-political theories, while relying on "hard" means (Clarkson et al., 2011). This is in contrast to both theories, according to which the more polluting companies, with worse environmental performance, would disclose more information, but through "soft" and unverifiable means (according to socio-political theories), while the better-performing organizations would use "hard" and verifiable means (according to voluntary disclosure theory). In this case, however, the worst-performing companies appear to disclose a lot by "hard" means. This raises significant concerns about the credibility of the information provided. Indeed, stakeholders (and especially investors attentive to sustainability issues) report that both the level (quantity) and nature (hard or soft) of disclosure may not be indicative of actual corporate performance.

This relates to the concepts of "organized hypocrisy" and "organizational facades" (Cho et al., 2015a), focusing on the consistency between disclosure and actions. The concept of "organized hypocrisy" is applicable to describe the differences that there may be between a company's disclosure, decisions, and actions: such divergences can ensure that the same company can manage the demands received from stakeholders more flexibly.

The term "organizational facades," on the other hand, indicates that there is no single facade of the organization; rather, there can be rational, progressive, and reputational facades that serve the enterprise to achieve its purposes in addition to social legitimacy. Employed together, these concepts allow discrepancies between disclosures and practices to generate benefits for a broad group of organization's stakeholders (Cho et al., 2015a). As emphasized by socio-political theories, corporate managers need to develop strategies to satisfy, or at least balance, stakeholders' expectations and not sever the implicit contract made with society (Barnett, 2007; Mitchell et al., 1997). If stakeholders are divided into groups (different but all necessary to ensure legitimacy for the company) and make conflicting informational demands, the company is called upon to devise strategies to satisfy (at least minimally) each of these demands.

Managerial responses in these situations often involve the use of organized hypocrisy: the corporate response in a scenario characterized by conflicting values, ideas, or people often involves inconsistencies between disclosure, decisions, and actions (Brunsson, 2007). As a result, organizations may develop ad hoc substructures to respond to the specific demands received, representing "organizational facades" of the organization. These are defined as a symbolic front set up by the organization to reassure its stakeholders about the legitimacy of the company and its management. They are, therefore, aimed at creating legitimacy in the eyes of stakeholders. They can be divided into three main types. First, rational facades are necessary to gain market legitimacy: they are employed to show an organization accommodating rational norms, i.e., the basic behavioral norms (e.g., making business decisions according to completed cost-benefit analyses). Second, progressive facades require considering progressive norms, thus applying the most cutting-edge managerial techniques and not deeming compliance with rational norms alone sufficient. Finally, reputational facades are used to display accounting and rhetorical symbols that critical stakeholders, such as analysts and the press, are seeking. They reflect particular corporate values and impact on an organization's image: they can reinforce a company's realistic objectives or mask business practices not accepted by certain stakeholders. Applying the aforementioned concepts to the issue of sustainability disclosure, Cho et al. (2015a) mark the need to place a company in the socio-institutional context in which it operates to trace the different, and even conflicting, expectations exerted on it by stakeholders because of their beliefs about sustainability. In such a situation, the facade of the first type (rational) is attributed to the current and growing logic of corporate sustainability, resulting from market pressures and capitalism, and provoking a continuous search for growth opportunities. Despite the prevalence of market logic, socio-environmental issues are emphasized in public debate and policy statements. This causes the development of facades of the second (progressive) type through the publication of voluntary sustainability reports, in which an attempt is made to provide answers to public concerns. This type of façade then becomes one designed to prioritize innovation and reform in the social and environmental spheres, based on the notion that technology can be used to remedy the negative impacts that societies have on the environment around them. However, it is important to point out that there is also a growing concern about whether the information communicated on these issues is merely symbolic because it contains omissions about negative items (e.g., environmentally unsustainable practices) and markedly emphasizes positive issues. Cho et al. (2015a) relate this approach to the use of facades of the third (reputational) type, having the goal of showing a socially and environmentally responsible organization that cares for the environment and the less well-off. The three types of facades make the differences between disclosed information and implemented practices in sustainability. There have also been ambiguous findings from several other studies conducted on the existence of a relationship between corporate financial performance and sustainability reporting. Al Hawaj and Buallay (2022) aimed to assess the existence of a relationship between economic performance (and its improvement) and non-financial reporting by analyzing specific industries. The purpose of their research was to understand whether or not the type of business conducted by companies could influence the relevance of sustainability reporting. The authors concluded that, in most of the economic sectors considered, the publication of non-financial reports has a positive impact and increases corporate value. In contrast, in the banking, IT, and telecommunications sectors, there is an inverse relationship between non-financial reporting and economic performance. In these three industries, burdens due to reporting ESG issues are found to be more costly than its economic benefits and return. These results, concerning most of the sectors in which the relationship between economic performance and sustainability reporting is positive, seem to confirm what has already been anticipated above, namely that non-financial reporting can be seen as an investment that has the capacity, through the satisfaction of stakeholder interests that follows, to increase corporate value and improve its performance. A very similar result has been reached by other authors. Berthelot et al. (2012) wanted to test whether or not the publication of NFI by companies has a positive effect on investors in the capital market. The authors concluded that the publication of sustainability reports is positively valued in the market, and they trace this result predominantly to one reason: non-financial reporting allows companies to publicize their commitment to ESG issues and improve their reputation, which also has positive effects on the financial market. This seems to confirm the idea that companies at the time they achieve positive results should communicate them so that they can benefit from the effects of those same results. Finally, De Klerk et al. (2015) get results aligned with the conclusions drawn by the other recalled studies, showing how the publication of non-financial reporting is positively correlated with corporate performance. They also go so far as to state that this link is much stronger for companies operating in those industries that may be more polluting or otherwise environmentally impactful than those that are not. The authors therefore suggest that companies should give the market information (both positive and negative) about their non-financial performance since they could benefit from it.

While numerous studies point to a positive link between sustainability reporting and corporate performance, there is a stream of literature that has not reached the same conclusion, as according to the latter there seems to be no real link between non-financial disclosure and corporate results. Specifically, Camodeca et al. (2018) argue that non-financial disclosure has a real impact from an economic point of view only when an independent organization is in charge of reviewing that document (i.e., assurance process), otherwise untrue information may have been indicated within it, for the sole purpose of improving the company's reputation and its positioning in the market. Qiu et al. (2016) sought to distinguish the effect generated by social reporting and the impact generated by environmental reporting. The authors concluded that the market welcomes corporate social reporting that can be used to gain approval from stakeholders and investors. In contrast, disclosure on environmental issues has a different effect: the authors were unable to demonstrate a true correlation between environmental reporting and corporate value. They therefore conclude that the only type of reporting that matters to the public is that on social issues, and companies should seek to extend it to maximize their economic benefit. The relationship between non-financial reporting and firm value thus appears to be quite dubious, with studies confirming its positive effect and studies that, otherwise, fail to detect a real link between the two items. While failing to detect the existence of a real impact from the financial point of view as a result of ESG disclosure, Guidry and Patten (2010) manage to show that companies, that disclose meaningful and relevant data in their non-financial reports, generate better reactions in the market than companies that do not pay much attention to the quality and accuracy of NFI reported. According to the authors, this is proof that if companies seek, through sustainability reporting, to increase their reputation in the market, they must pay close attention to the quality of the information disseminated and how it is communicated.

Analysis of Information Quantity and Quality in Sustainability Reporting

A substantial part of the literature analyzes sustainability disclosures by organizations (mainly private ones). There are several measurement techniques used in so-called content analysis. Joseph and Taplin (2011) evaluated two types of measurements that can be conducted in this regard: information abundance (to be understood as the volume of information provided from a purely quantitative point of view) and the occurrence of disclosures, which can be measured through an appropriate disclosure index. Information volume is generally measured by counting the number of words, sentences, or pages devoted to sustainability information, while the occurrence of disclosures is measured by using a kind of checklist to check the presence of information regarding a set of predetermined topics, assessing the variety of NFI provided. In general, more importance is accorded to the findings of qualitative analysis than quantitative one, but both have weaknesses. For example, checking the volume of disclosures can lead to the problem of double-counting an item when it is reported in two different parts of the text and, therefore, will be counted twice. In contrast, qualitative information analysis avoids this problem, but omits consideration of the disclosure volume, as it is sufficient if there is, for example, only one sentence to attest to the presence of that particular topic or subject matter. Therefore, the adoption of one type of analysis, rather than the other, can lead to potentially different conclusions.

With reference to *quantitative* analysis, a substantial number of studies confirm the increasing trend in NFI volume in recent years concerning both the number of non-financial reports and the number of pages of such reports (both annual and stand-alone) dealing with sustainability issues (Rimmel & Jonäll, 2013; Stolowy & Paugam, 2018). The literature shows that this quantitative increase is affected by both regulatory and stakeholder pressures. With reference to the former, for example, Weerathunga et al. (2020) analyzed whether the convergence and adoption of IFRS affected CSR disclosures of Indian listed companies. The objective was to examine if the adoption of IFRS leads to an increase in non-financial disclosure. As a unit of measure, the authors employed the number of words, and the results show that companies adopting the IFRS significantly increased their level of CSR disclosure with reference to all its dimensions, concerning employees, human rights, environment, society, and community.

With reference to *qualitative* analysis, non-financial disclosures were examined through the use of indices designed to determine the variety of report content. Sustainability reporting should be complete (containing information on all relevant issues, both positive and negative), accurate (allowing stakeholders to make their assessments in an appropriate and informed manner), and reliable (information content can be verified through the evidence provided). Generally, quality assessment can be broken down into at least three levels (Cormier & Gordon, 2001). It is precisely related to reliability (Comyns et al., 2013) by classifying information into three groups: "search" (for information that can be easily verified by the recipient and for which higher quality is expected), "experience" (for information that can be verified in the future by the recipient, assuming that its quality improves over time), and "credence" (for information that will never be verifiable and, therefore, for which reduced permanent quality is expected). Always considering a three-level articulation, information quality is linked to the completeness of sustainability disclosure. The highest qualitative level (i.e., the one corresponding to greater informational completeness) includes quantitative narratives (i.e., evaluations also enriched by numerical parameters about sustainability issues); the intermediate qualitative level allows the recipient of the report to benefit from the description (detailed though not numerical) of sustainability issues; finally, the low qualitative level of disclosure involves only general mentions of a topic (Agostini & Costa, 2018; Agostini et al., 2022; Costa & Agostini, 2016; Michelon et al., 2015).

The analysis of the informational quality of reports is of particular importance considering also that an authoritative part of the literature has shown that there is a positive relationship between this quality and the value attributed to the company (Barth et al., 2017).

However, the quality of sustainability reports has often been the subject of criticism (Gray, 2010). In particular, the quality of disclosures can be hindered by information asymmetry between issuing companies and information recipients. This occurs because only the company has at its disposal full knowledge of sustainability issues relevant (in both positive and negative senses) to its business. This implies that the recipients of the report may not be able to fully understand the actual quality of the information disclosed (even voluntarily) by the companies and, therefore, the companies may not be incentivized in turn to increase the quality of the information provided, as there is no assurance of understanding and reliance by the recipients (Healy & Palepu, 2001; Kulkarni, 2000; Plumlee et al., 2015).

Multiple studies have examined the information quality concerning environmental issues given the increasing general focus on the topic (Adler et al., 2017). Several studies in this regard have concluded that while it varies from company to company, in general, there is an increase in information provided on the topic in question over the years, especially by larger companies, probably because they have more funds and resources at their disposal. However, this does not correspond to more detailed and accurate disclosures about the initiatives and impacts of the activity (Adler et al., 2018). Indeed, even a substantial increase in the number of companies engaged in sustainability reporting does not automatically correlate with a higher quality of information in terms of completeness, accuracy, and transparency, suggesting that companies often intend to maintain their legitimacy without increasing the quality of their disclosures and calling out the need for appropriate regulation (Barkemeyer et al., 2015; Comyns & Figge, 2015).

Melloni et al. (2017) focus on reporting features that are related to quality: conciseness, completeness, and balance. They are analyzed within the integrated reports of companies pioneering the adoption of such a model. Specifically, to measure the completeness of information, the authors assessed whether certain topics were covered and employed the "Bloomberg ESG disclosure scores" (Lai et al., 2017). Moreover, they focused on the tone of the disclosures, specifically the optimism shown to them. Overall, the results of this study state that, for companies with poor performance, supporting the legitimacy theory, integrated reports tend to be longer, less concise, more complex, and convey a lower level of ESG information (less complete reports). This indicates a "manipulative" and strategic purpose in the provision of NFI. This seems to emphasize the need for appropriate regulations that can create additional pressures and economic incentives toward greater transparency and informational qualities (Christ et al., 2019). Haque et al. (2016), considering a climate change disclosure index, indicate that most of the items in the index investigated were not disclosed by companies. They highlight a reluctance of the surveyed companies toward disclosure transparency, preferring to provide information about economic-financial issues rather than issues responsive to community needs.

Most of the relevant literature has focused on the private sector: there is a very limited number of contributions on the topic examined here concerning the public sector (Farneti & Guthrie, 2009). While the number of companies engaged in sustainability reporting has gradually increased since the 1990s, in the public sector this practice is still in its infancy and is still characterized by slow development, coexistence of international and national standards, and comparative difficulties (Moggi, 2019). The available studies that focus on the public sector highlight the wide use of GRI guidelines, although in a fragmented way and using only some of the guidance (Guthrie & Farneti, 2008). Regarding the quality of public organizations' sustainability reports and disclosures, some studies point to shortcomings in consistency, completeness, transparency, and control mechanisms, thus undermining the very credibility of such reports (Chiba et al., 2018).

Assurance of Sustainability Reports

As emphasized in the previous sections, the increased adoption of sustainability reporting allows stakeholders to have much more information regarding the organization, going beyond the data retrieved within the annual financial statements. However, the reasons why companies opt for the preparation of non-financial reports are manifold. These could be related to the willingness to disclose actual positive sustainability performance, or they could be a way to improve corporate reputation especially when the actual results achieved are not so excellent, as emphasized above by the recalled legitimacy theory. These concerns toward the content and quality of sustainability reports have led in recent years to an increase in the spread of auditing practices, implemented by third and independent parties. Without an effective assessment of the information disclosed, it becomes difficult to understand if the actual performance of companies is based on what is reported (Braam et al., 2016; Cuadrado-Ballesteros et al., 2017; Martínez-Ferrero and García-Sánchez, 2017).

Kim et al. (2019) affirm that assurance allows to improve the quality and reliability of the disclosure, emphasizing that NFI is clearer and more transparent when an independent organization evaluates its content. Therefore, companies that want to make sure that their non-financial disclosures are perceived as more reliable and of higher quality should have their reports audited. This allows for a reduction in the information asymmetries between stakeholders and the company itself (Cuadrado-Ballesteros et al., 2017). Braam et al. (2016) assess the relevance of assurance concerning sustainability reports, seeking to understand when the latter, in a non-mandatory context, was most used. The authors consider a sample of 100 Dutch companies and conclude that the companies that make the most use of the assurance of their non-financial statements are those that operate in sectors where there is a high environmental impact (as written concerning the use of voluntary reporting). They also note a positive relationship between the judgment drafted by independent professionals and the transparency of NFI reported. Additionally, they point out how the companies that audit their non-financial statements are the same ones that present the most information regarding their performance. García-Sánchez et al. (2022) highlight how the usefulness of assurance manifests itself not only when companies provide information that does not fully correspond to the reality of their business (the practice of greenwashing), but also when they fail to communicate the information clearly, making their performance appear worse than it is. The authors reiterate that a mismatch between what is achieved and what is reported can also be harmful in the latter case, as it risks making the company's practices appear inadequate concerning the needs of stakeholders. Martínez-Ferrero and García-Sánchez (2017) highlight an additional benefit that comes from reviewing sustainability reports. Indeed, the authors highlight how the assurance, through its ability to reduce both uncertainty (concerning the reported information) and perceived risk (by stakeholders), allows for a decrease in the cost of capital for companies. Based on a sample of 1410 international companies, they conclude that the assurance of sustainability reports brings with it an increase in the company's reputation, with a consequent reduction in the risk perceived by investors; all of which is reflected in a reduction in the return expected by investors on their investments. After focusing on the positive effects that can result from reviewing sustainability reports, the authors (García-Sánchez et al., 2022) point out that for the assurance that is carried out by experts to be relevant, they need to have a good understanding of the goals and practices that the company implements concerning ESG issues, to give an evaluation that is consistent with the targets that the company aims to achieve.

About the assurance of sustainability reporting, mixed results emerge in the literature as well. Michelon et al. (2015) do not retrieve greater quality and quantity of information in audited reports, concluding that such assurance does not play a substantive role, but is more symbolic and appearance, in line with legitimacy theory. This is closely related to the so-called expectation gap in auditing, which represents the difference between what stakeholders expect from assurance and what assurance provides. Therefore, it corresponds to the difference between the perception and the reality of assurance. It may be mainly due to a lack of understanding on the part of the stakeholders or a legitimate concern. Stakeholders' expectations concerning assurance may grow depending on what auditors can perform and achieve. Building on these findings, Maroun (2019) highlights that of paramount importance is also the professionalism of those who carry out the report assurance. From this point of view, this research, among others, points out that the evaluations carried out by assurance companies falling within the group of the *Big 4* (Deloitte, PricewaterhouseCoopers, Ernst&Young, and KPMG) turn out to be of higher quality, given the high level of professionalism in carrying out the audit activities. Similar findings were also identified by Boiral et al. (2020). These authors highlight that, when report assurance is carried out by independent parties not belonging to large consulting firms, the controls are perceived as less reliable, mainly as a consequence of the few resources available to the experts. Conversely, when the same activity is carried out by one of the *Big 4* or parties related to them, this is perceived as more serious and accurate.

Several authors, including Cuadrado-Ballesteros et al. (2017) and Maroun (2019), also highlight how different levels of control can be implemented when checking the quality of the reports produced. Indeed, consultants involved in auditing activities can implement *limited assurance* (i.e., moderate) or *reasonable assurance* (i.e., high and accurate). The difference between the two lies precisely in the level of depth and precision with which the control activities are carried out.

Limited assurance presents an acceptable, but not very specific, level of control, and the conclusion the auditor arrives at is expressed negatively, confirming that he or she has found no evidence of noncompliance with reporting standards or the presentation of untrue information. On the other hand, reasonable assurance is much more precise, while still not giving absolute certainty, and the conclusion is expressed positively, thus affirming the conformity of what is stated with reality. According to the authors, the level of depth in the assessment of NFI also influences the perception of stakeholders, who, when reports are assessed with greater accuracy (reasonable assurance), attribute greater reliability to the data contained in the reports themselves.

The role of assurance lends greater reliability and credibility to the reports reviewed. Thorne et al. (2014), in their literature review, noted the existence of the "*Assurance theory*", according to which sustainability reports hold greater reliability when reviewed by a professional accountant. In addition, the inclusion of an assurance statement within sustainability reports increases the perception of credibility in the eyes of stakeholders (Fuhrmann et al., 2017). In addition, auditors provide recommendations to companies regarding materiality assessment and stakeholder engagement, influencing and facilitating the selection of significantly appropriate items on which to focus disclosures (Adler et al., 2017). Finally, the topic of auditing sustainability disclosures was also addressed by Farooq and De Villiers (2020), through 50 interviews with assurance service providers and managers responsible for sustainability reporting in Australia and New Zealand. Both groups surveyed demonstrated an understanding of the importance of assurance engagement, emphasizing the need for the engagement to be carried out by an independent entity to increase the reliability and fairness of the audit. To increase the quality of sustainability reports, especially regarding the reliability of reported NFI, the study highlights that there are two practices to be implemented in combination: the pre-assignment consultation and the actual audit. However, despite these practices and the guidance provided, companies remain reluctant to disclose relevant information with negative implications.

4.4   Concluding Remarks

The role and motivation of sustainability reporting have been extensively studied through the application of different theoretical frameworks, based on which the examined studies also hypothesized and verified the relation between non-financial disclosure and actual corporate performance. Furthermore, the relevance of conducting an analysis that investigates the quality and completeness of NFI contained in sustainability reports, not merely dwelling on quantitative expectations (such as the number of words, pages, or sentences) has emerged. However, the main qualitative analyses implemented in the examined literature are on rather specific topics (such as environmental issues, biodiversity, climate change, and modern slavery). Thus, there is a lack in the reviewed literature of guide-lines that can be applied generally to support the different forms that sustainability reporting can take, given also the different frameworks that companies can adopt to prepare it. In addition, almost all of the analysis was conducted having voluntary disclosures as a reference, only partially addressing the shift from a voluntary to a mandatory reporting approach. In this regard, a stream of literature states that the mere provision of regulatory requirements is not sufficient to increase the quality of NFI provided by companies. There emerges the need for a change in corporate culture, so that NFI provided to stakeholders is not perceived merely as an obligation, but as an opportunity to demonstrate and increase the commitment regarding sustainability and performance issues. The quality of the information provided is also of high importance from the point of view of financial impact. While there is no unambiguous demonstration of a positive relationship between sustainability reporting and financial performance, most studies agree that quality disclosures nevertheless have a positive effect vis-à-vis the market. Finally, it is pointed out that the assurance of sustainability reports, by third-party and independent experts, is relevant in improving stakeholders' perception of NFI disclosed, to make the information more reliable and transparent.

REFERENCES

Accountability. (2013). *Redefining materiality II: Why it matters, who's involved, and what it means for corporate leaders and boards.* https://lifegateedu.it/wp-content/uploads/2021/09/AA_Materiality_Report_Aug2013-FINAL_compressed.pdf

Adams, C. A. (2002). Internal organisational factors influencing corporate social and ethical reporting: Beyond current theorising. *Accounting, Auditing & Accountability Journal, 15*(2), 223–250.

Adams, C. A., & Larrinaga-González, C. (2007). Engaging with organisations in pursuit of improved sustainability accounting and performance. *Accounting, Auditing & Accountability Journal, 20*(3), 333–355.

Adams, C. A., Potter, B., Singh, P. J., & York, J. (2016). Exploring the implications of integrated reporting for social investment (disclosures). *The British Accounting Review, 48*(3), 283–296.

Adler, R., Mansi, M., Pandey, R., & Stringer, C. (2017). United Nations Decade on biodiversity: A study of the reporting practices of the Australian mining industry. *Accounting, Auditing & Accountability Journal, 30*(8), 1711–1745.

Adler, R., Mansi, M., & Pandey, R. (2018). Biodiversity and threatened species reporting by the top Fortune Global companies. *Accounting, Auditing & Accountability Journal, 31*(3), 787–825.

Agostini, M., Costa, E., & Korca, B. (2022). Non-financial disclosure and corporate financial performance under directive 2014/95/EU: Evidence from Italian listed companies. *Accounting in Europe, 19*(1), 78–109.

Agostini, M., & Costa, E. (2018). Financial and sustainability reporting: An empirical investigation of their relationship in the Italian context. *Sustainability and Social Responsibility: Regulation and Reporting*, 411–441.

Al Hawaj, A. Y., & Buallay, A. M. (2022). A worldwide sectorial analysis of sustainability reporting and its impact on firm performance. *Journal of Sustainable Finance & Investment, 12*(1), 62–86.

Aureli, S., Magnaghi, E., & Salvatori, F. (2019). The role of existing regulation and discretion in harmonising non-financial disclosure. *Accounting in Europe, 16*(3), 290–312.

Barbu, E. M., Ionescu-Feleagă, L., & Ferrat, Y. (2022). The evolution of environmental reporting in Europe: The role of financial and non-financial regulation. *The International Journal of Accounting, 57*(2), 2250008.

Barkemeyer, R., Preuss, L., & Lee, L. (2015, December). Corporate reporting on corruption: An international comparison. *Accounting Forum, 39*(4), 349–365.

Barnett, M. L. (2007). Stakeholder influence capacity and the variability of financial returns to corporate social responsibility. *Academy of Management Review, 32*(3), 794–816.

Barone, E., Ranamagar, N., & Solomon, J. F. (2013, September). A Habermasian model of stakeholder (non) engagement and corporate (ir) responsibility reporting. *Accounting Forum, 37*(3), 163–181.

Barth, M. E., Cahan, S. F., Chen, L., & Venter, E. R. (2017). The economic consequences associated with integrated report quality: Capital market and real effects. *Accounting, Organizations and Society, 62*, 43–64.

Bebbington, J., Higgins, C., & Frame, B. (2009). Initiating sustainable development reporting: Evidence from New Zealand. *Accounting, Auditing & Accountability Journal, 22*(4), 588–625.

Bebbington, J., Kirk, E., & Larrinaga, C. (2012). The production of normativity: A comparison of reporting regimes in Spain and the UK. *Accounting, Organizations and Society, 37*(2), 78–94.

Bebbington, J., & Larrinaga, C. (2014). Accounting and sustainable development: An exploration. *Accounting, Organizations and Society, 39*(6), 395–413.

Bellucci, M., Acuti, D., Simoni, L., & Manetti, G. (2021). Restoring an eroded legitimacy: The adaptation of nonfinancial disclosure after a scandal and the risk of hypocrisy. *Accounting, Auditing & Accountability Journal, 34*(9), 195–217.

Bellucci, M., Simoni, L., Acuti, D., & Manetti, G. (2019). Stakeholder engagement and dialogic accounting: Empirical evidence in sustainability reporting. *Accounting, Auditing & Accountability Journal, 32*(5), 1467–1499.

Berthelot, S., Coulmont, M., & Serret, V. (2012). Do investors value sustainability reports? A Canadian study. *Corporate Social Responsibility and Environmental Management, 19*(6), 355–363.

Beske, F., Haustein, E., & Lorson, P. C. (2020). Materiality analysis in sustainability and integrated reports. *Sustainability Accounting, Management and Policy Journal, 11*(1), 162–186.

Biondi, L., Dumay, J., & Monciardini, D. (2020). Using the international integrated reporting framework to comply with EU directive 2014/95/EU: Can we afford another reporting façade? *Meditari Accountancy Research, 28*(5), 889–914.

Blacconiere, W. G., & Patten, D. M. (1994). Environmental disclosures, regulatory costs, and changes in firm value. *Journal of Accounting and Economics, 18*(3), 357–377.

Bold, F. (2017), Compliance and reporting under the EU non-financial reporting directive: Requirements and opportunities, Czech Republic, Brussels, Brno.

Boiral, O., Heras-Saizarbitoria, I., & Brotherton, M. C. (2020). Professionalizing the assurance of sustainability reports: The auditors' perspective. *Accounting, Auditing & Accountability Journal, 33*(2), 309–334.

Bouten, L., & Everaert, P. (2015). Social and environmental reporting in Belgium: 'Pour vivre heureux, vivons cachés.' *Critical Perspectives on Accounting, 33*, 24–43.

Bouten, L., Everaert, P., Van Liedekerke, L., De Moor, L., & Christiaens, J. (2011). Corporate social responsibility reporting: A comprehensive picture? *Accounting Forum, 35*(3), 187–204.

Bowen, R. M., Davis, A. K., & Matsumoto, D. A. (2005). Emphasis on pro forma versus GAAP earnings in quarterly press releases: Determinants, SEC intervention, and market reactions. *The Accounting Review, 80*(4), 1011–1038.

Braam, G. J., De Weerd, L. U., Hauck, M., & Huijbregts, M. A. (2016). Determinants of corporate environmental reporting: The importance of environmental performance and assurance. *Journal of Cleaner Production, 129*, 724–734.

Brammer, S., & Pavelin, S. (2008). Factors influencing the quality of corporate environmental disclosure. *Business Strategy and the Environment, 17*(2), 120–136.

Brown, J. (2009). Democracy, sustainability and dialogic accounting technologies: Taking pluralism seriously. *Critical Perspectives on Accounting, 20*(3), 313–342.

Brown, J., & Dillard, J. (2013). Agonizing over engagement: SEA and the "death of environmentalism" debates. *Critical Perspectives on Accounting*, *24*(1), 1–18.

Brown, J., & Dillard, J. (2014). Integrated reporting: On the need for broadening out and opening up. *Accounting, Auditing & Accountability Journal*, *27*(7), 1120–1156.

Brown, J., & Dillard, J. (2015). Dialogic accountings for stakeholders: On opening up and closing down participatory governance. *Journal of Management Studies*, *52*(7), 961–985.

Brunsson, N. (2007). *The consequences of decision-making*. Oxford University Press.

Buhr, N. (2002). A structuration view on the initiation of environmental reports. *Critical Perspectives on Accounting*, *13*(1), 17–38.

Buhr, N. (1998). Environmental performance, legislation and annual report disclosure: The case of acid rain and Falconbridge. *Accounting, Auditing & Accountability Journal*, *11*(2), 163–190.

Camodeca, R., Almici, A., & Sagliaschi, U. (2018). Sustainability disclosure in integrated reporting: Does it matter to investors? A cheap talk approach. *Sustainability*, *10*(12), 4393.

Cerbone, D., & Maroun, W. (2020). Materiality in an integrated reporting setting: Insights using an institutional logics framework. *The British Accounting Review*, *52*(3), 100876.

Chaidali, P., & Jones, M. J. (2017). It's a matter of trust: Exploring the perceptions of integrated reporting preparers. *Critical Perspectives on Accounting*, *48*, 1–20.

Chiba, S., Talbot, D., & Boiral, O. (2018). Sustainability adrift: An evaluation of the credibility of sustainability information disclosed by public organizations. *Accounting Forum*, *42*(4), 328–340.

Cho, C. H., Laine, M., Roberts, R. W., & Rodrigue, M. (2015a). Organized hypocrisy, organizational façades, and sustainability reporting. *Accounting, Organizations and Society*, *40*, 78–94.

Cho, C. H., Michelon, G., & Patten, D. M. (2012). Impression management in sustainability reports: An empirical investigation of the use of graphs. *Accounting and the Public Interest*, *12*(1), 16–37.

Cho, C. H., Michelon, G., Patten, D. M., & Roberts, R. W. (2015b). CSR disclosure: The more things change...? *Accounting, Auditing & Accountability Journal*, *28*(1), 14–35.

Cho, C. H., & Patten, D. M. (2013). Green accounting: Reflections from a CSR and environmental disclosure perspective. *Critical Perspectives on Accounting*, *24*(6), 443–447.

Christ, K. L., Rao, K. K., & Burritt, R. L. (2019). Accounting for modern slavery: An analysis of Australian listed company disclosures. *Accounting, Auditing & Accountability Journal, 32*(3), 836–865.

Clarkson, P. M., Overell, M. B., & Chapple, L. (2011). Environmental reporting and its relation to corporate environmental performance. *Abacus, 47*(1), 27–60.

Clarkson, P. M., Li, Y., Richardson, G. D., & Vasvari, F. P. (2008). Revisiting the relation between environmental performance and environmental disclosure: An empirical analysis. *Accounting, Organizations and Society, 33*(4–5), 303–327.

Comyns, B., & Figge, F. (2015). Greenhouse gas reporting quality in the oil and gas industry: A longitudinal study using the typology of "search", "experience" and "credence" information. *Accounting, Auditing & Accountability Journal, 28*(3), 403–433.

Comyns, B., Figge, F., Hahn, T., & Barkemeyer, R. (2013). Sustainability reporting: The role of "Search", "Experience" and "Credence" information. *Accounting Forum, 37*(3), 231–243).

Cormier, D., & Gordon, I. M. (2001). An examination of social and environmental reporting strategies. *Accounting, Auditing & Accountability Journal, 14*(5), 587–617.

Costa, E., & Agostini, M. (2016). Mandatory disclosure about environmental and employee matters in the reports of Italian-listed corporate groups. *Social and Environmental Accountability Journal, 36*(1), 10–33.

Crane, A., & Ruebottom, T. (2011). Stakeholder theory and social identity: Rethinking stakeholder identification. *Journal of Business Ethics, 102*, 77–87.

Cuadrado-Ballesteros, B., Martínez-Ferrero, J., & García-Sánchez, I. M. (2017). Mitigating information asymmetry through sustainability assurance: The role of accountants and levels of assurance. *International Business Review, 26*(6), 1141–1156.

De Klerk, M., De Villiers, C., & Van Staden, C. (2015). The influence of corporate social responsibility disclosure on share prices: Evidence from the United Kingdom. *Pacific Accounting Review, 27*(2), 208–228.

De Villiers, C., & Alexander, D. (2014). The institutionalisation of corporate social responsibility reporting. *The British Accounting Review, 46*(2), 198–212.

Deegan, C., Rankin, M., & Tobin, J. (2002). An examination of the corporate social and environmental disclosures of BHP from 1983–1997: A test of legitimacy theory. *Accounting, Auditing & Accountability Journal, 15*(3), 312–343.

Deegan, C. (2002). Introduction: The legitimising effect of social and environmental disclosures–a theoretical foundation. *Accounting, auditing & accountability journal, 15*(3), 282–311.

Delbard, O. (2008). CSR legislation in France and the European regulatory paradox: An analysis of EU CSR policy and sustainability reporting practice. *Corporate Governance, 8*, 397–405.

Dillard, J., & Roslender, R. (2011). Taking pluralism seriously: Embedded moralities in management accounting and control systems. *Critical Perspectives on Accounting, 22*(2), 135–147.

Dilling, P. F., & Harris, P. (2018). Reporting on long-term value creation by Canadian companies: A longitudinal assessment. *Journal of Cleaner Production, 191*, 350–360.

DiMaggio, P. J., & Powell, W. W. (1983). The iron cage revisited: Institutional isomorphism and collective rationality in organizational fields. *American Sociological Review*, 147–160.

Dye, R. A. (1985). Disclosure of nonproprietary information. *Journal of Accounting Research*, 123–145.

Doshi, A. R., Dowell, G. W., & Toffel, M. W. (2013). How firms respond to mandatory information disclosure. *Strategic Management Journal, 34*(10), 1209–1231.

Dumay, J., La Torre, M., & Farneti, F. (2019). Developing trust through stewardship: Implications for intellectual capital, integrated reporting, and the EU directive 2014/95/EU. *Journal of Intellectual Capital, 20*(1), 11–39.

Dumitru, M., Dyduch, J., Gușe, R., & Krasodomska, J. (2017). Corporate reporting practices in Poland and Romania–an ex-ante study to the new non-financial reporting European directive. *Accounting in Europe, 14*(3), 279–304.

Eccles, R. G., Krzus, M. P., Rogers, J., & Serafeim, G. (2012). The need for sector-specific materiality and sustainability reporting standards. *Journal of Applied Corporate Finance, 24*(2), 65–71.

Erkens, M., Paugam, L., & Stolowy, H. (2015). Non-financial information: State of the art and research perspectives based on a bibliometric study. *Comptabilité-Contrôle-Audit, 21*(3), 15–92.

Farneti, F., & Guthrie, J. (2009, June). Sustainability reporting by Australian public sector organisations: Why they report. *Accounting Forum, 33*(2), 89–98.

Farooq, M. B., & De Villiers, C. (2019). Understanding how managers institutionalise sustainability reporting: Evidence from Australia and New Zealand. *Accounting, Auditing & Accountability Journal, 32*(5), 1240–1269.

Farooq, M. B., & De Villiers. (2020). How sustainability assurance engagement scopes are determined, and its impact on capture and credibility enhancement. *Accounting, Auditing & Accountability Journal, 33*(2), 417–445.

Fasan, M., & Mio, C. (2017). Fostering stakeholder engagement: The role of materiality disclosure in integrated reporting. *Business Strategy and the Environment, 26*(3), 288–305.

Freeman, R.E. (1984). *Strategic management: A stakeholder approach.* Prentice-Hall, Englewood Cliffs.

Frost, G., Jones, S., Loftus, J., & Van Der Laan, S. (2005). A survey of sustainability reporting practices of Australian reporting entities. *Australian Accounting Review, 15*(35), 89–96.

Fuhrmann, S., Ott, C., Looks, E., & Guenther, T. W. (2017). The contents of assurance statements for sustainability reports and information asymmetry. *Accounting and Business Research, 47*(4), 369–400.

García-Sánchez, I. M., Hussain, N., Aibar-Guzmán, C., & Aibar-Guzmán, B. (2022). Assurance of corporate social responsibility reports: Does it reduce decoupling practices? *Business Ethics, the Environment & Responsibility, 31*(1), 118–138.

Gray, R. (2000). Current developments and trends in social and environmental auditing, reporting and attestation: A review and comment. *International Journal of Auditing, 4*(3), 247–268.

Gray, R. (2002). The social accounting project and accounting organizations and society privileging engagement, imaginings, new accountings and pragmatism over critique? *Accounting, Organizations and Society, 27*(7), 687–708.

Gray, R. (2010). Is accounting for sustainability actually accounting for sustainability... and how would we know? An exploration of narratives of organisations and the planet. *Accounting, Organizations and Society, 35*(1), 47–62.

Gray, R., Dey, C., Owen, D., Evans, R., & Zadek, S. (1997). Struggling with the praxis of social accounting: Stakeholders, accountability, audits and procedures. *Accounting, Auditing & Accountability Journal, 10*(3), 325–364.

Gray, R., Kouhy, R., & Lavers, S. (1995). Corporate social and environmental reporting: A review of the literature and a longitudinal study of UK disclosure. *Accounting, Auditing & Accountability Journal, 8*(2), 47–77.

GRI. (2013). G4 sustainability reporting guidelines: reporting principles and standard disclosures. *Global Reporting Initiative*, Amsterdam. Retrieved from: https://respect.international/g4-sustainability-reporting-guidelines-reporting-principles-and-standard-disclosures/.

Gray, I., & Manson, S. (2008). *The audit process.* Thomson Learning.

Guidry, R. P., & Patten, D. M. (2010). Market reactions to the first-time issuance of corporate sustainability reports: Evidence that quality matters. *Sustainability Accounting, Management and Policy Journal, 1*(1), 33–50.

Guthrie, J., Cuganesan, S., & Ward, L. (2008, March). Industry specific social and environmental reporting: The Australian Food and Beverage Industry. *Accounting forum, 32*(1), 1–15.

Guthrie, J., & Farneti, F. (2008). GRI sustainability reporting by Australian public sector organizations. *Public Money and Management, 28*(6), 361–366.

Haque, S., Deegan, C., & Inglis, R. (2016). Demand for, and impediments to, the disclosure of information about climate change-related corporate governance practices. *Accounting and Business Research, 46*(6), 620–664.

Hrasky, S. (2012). Carbon footprints and legitimation strategies: Symbolism or action? *Accounting, Auditing & Accountability Journal, 25*(1), 174–198.

Healy, P. M., & Palepu, K. G. (2001). Information asymmetry, corporate disclosure, and the capital markets: A review of the empirical disclosure literature. *Journal of Accounting and Economics, 31*(1–3), 405–440.

Hopwood, A. G. (2009). Accounting and the environment. *Accounting, Organizations and Society, 34*(3–4), 433–439.

International Integrated Reporting Council (IIRC). (2013). *Materiality background paper for IR*. https://www.integratedreporting.org/wp-content/uploads/2013/03/IR-Background-Paper-Materiality.pdf

Jones, M. J. (2010, June). Accounting for the environment: Towards a theoretical perspective for environmental accounting and reporting. *Accounting Forum, 34*(2), 123–138.

Joseph, C., & Taplin, R. (2011, March). The measurement of sustainability disclosure: Abundance versus occurrence. *Accounting Forum, 35*(1), 19–31.

Kim, J., Cho, K., & Park, C. K. (2019). Does CSR assurance affect the relationship between CSR performance and financial performance? *Sustainability, 11*(20), 5682.

KPMG. (2017). *The KPMG survey of corporate responsibility reporting 2017*. https://assets.kpmg/content/dam/kpmg/xx/pdf/2017/10/kpmg-survey-of-corporate-responsibility-reporting-2017.pdf

KPMG. (2022). *KPMG global survey of sustainability reporting 2022*. https://kpmg.com/it/it/home/insights/2022/12/kpmg-global-survey-of-sustainability-reporting-2022.html

Kulkarni, S. P. (2000). Environmental ethics and information asymmetry among organizational stakeholders. *Journal of Business Ethics, 27*, 215–228.

La Torre, M., Sabelfeld, S., Blomkvist, M., Tarquinio, L., & Dumay, J. (2018). Harmonising non-financial reporting regulation in Europe: Practical forces and projections for future research. *Meditari Accountancy Research, 26*(4), 598–621.

Lai, A., Melloni, G., & Stacchezzini, R. (2017). What does materiality mean to integrated reporting preparers? An empirical exploration. *Meditari Accountancy Research, 25*(4), 533–552.

Leong, S., & Hazelton, J. (2019). Under what conditions is mandatory disclosure most likely to cause organisational change? *Accounting, Auditing & Accountability Journal, 32*(3), 811–835.

Lepoutre, J. M., & Valente, M. (2012). Fools breaking out: The role of symbolic and material immunity in explaining institutional nonconformity. *Academy of Management Journal, 55*(2), 285–313.

Lindblom, C. K. (1994). The implications of organizational legitimacy for corporate social performance and disclosure. In *Critical Perspectives on Accounting Conference, New York*.

Lin-Hi, N., & Müller, K. (2013). The CSR bottom line: Preventing corporate social irresponsibility. *Journal of Business Research, 66*(10), 1928–1936.

Littell, J. H., Corcoran, J., & Pillai, V. (2008). *Systematic reviews and meta-analysis*. Oxford.

Lounsbury, M. (2008). Institutional rationality and practice variation: New directions in the institutional analysis of practice. *Accounting, Organizations and Society, 33*(4–5), 349–361.

Luque-Vilchez, M., & Larrinaga, C. (2016). Reporting models do not translate well: Failing to regulate CSR reporting in Spain. *Social and Environmental Accountability Journal, 36*(1), 56–75.

Magness, V. (2006). Strategic posture, financial performance and environmental disclosure: An empirical test of legitimacy theory. *Accounting, Auditing & Accountability Journal, 19*(4), 540–563.

Mäkelä, H., & Näsi, S. (2010). Social responsibilities of MNCs in downsizing operations: A Finnish forest sector case analysed from the stakeholder, social contract and legitimacy theory point of view. *Accounting, Auditing & Accountability Journal, 23*(2), 149–174.

Malsch, B. (2013). Politicizing the expertise of the accounting industry in the realm of corporate social responsibility. *Accounting, Organizations and Society, 38*(2), 149–168.

Manetti, G., & Bellucci, M. (2016). The use of social media for engaging stakeholders in sustainability reporting. *Accounting, Auditing & Accountability Journal, 29*(6), 985–1011.

Maroun, W. (2019). Does external assurance contribute to higher quality integrated reports? *Journal of Accounting and Public Policy, 38*(4), 106670.

Martin-Sardesai, A., & Guthrie, J. (2019). Social report innovation: Evidence from a major Italian bank 2007–2012. *Meditari Accountancy Research, 28*(1), 72–88.

Martínez-Ferrero, J., & García-Sánchez, I. M. (2017). Sustainability assurance and cost of capital: Does assurance impact on credibility of corporate social responsibility information? *Business Ethics: A European Review, 26*(3), 223–239.

Massaro, M., Dumay, J., & Guthrie, J. (2016). On the shoulders of giants: Undertaking a structured literature review in accounting. *Accounting, Auditing & Accountability Journal, 29*(5), 767–801.

Melloni, G., Caglio, A., & Perego, P. (2017). Saying more with less? Disclosure conciseness, completeness and balance in Integrated Reports. *Journal of Accounting and Public Policy, 36*(3), 220–238.

Merkl-Davies, D. M., & Brennan, N. M. (2007). Discretionary disclosure strategies in corporate narratives: Incremental information or impression management? *Journal of Accounting Literature, 27*, 116–196.

Meyer, J. W., & Rowan, B. (1991). Institutionalized organizations: formal structure as myth and ceremony. *The new institutionalism in organizational analysis*, 41–62.

Michelon, G., Pilonato, S., & Ricceri, F. (2015). CSR reporting practices and the quality of disclosure: An empirical analysis. *Critical Perspectives on Accounting, 33*, 59–78.

Miles, S. (2017). Stakeholder theory classification: A theoretical and empirical evaluation of definitions. *Journal of Business Ethics, 142*, 437–459.

Miles, S., & Ringham, K. (2020). The boundary of sustainability reporting: Evidence from the FTSE100. *Accounting, Auditing & Accountability Journal, 33*(2), 357–390.

Milne, M., & Gray, R. H. (2007). Future prospects for sustainability reporting. In *Sustainability accounting and accountability* (pp. 184–208). Routledge Taylor & Francis Group.

Milne, M. J., & Gray, R. (2013). W (h) ither ecology? The triple bottom line, the global reporting initiative, and corporate sustainability reporting. *Journal of Business Ethics, 118*, 13–29.

Milne, M. J., Tregidga, H., & Walton, S. (2009). Words not actions! The ideological role of sustainable development reporting. *Accounting, Auditing & Accountability Journal, 22*(8), 1211–1257.

Mitchell, R. K., Agle, B. R., & Wood, D. J. (1997). Toward a theory of stakeholder identification and salience: Defining the principle of who and what really counts. *Academy of Management Review, 22*(4), 853–886.

Moggi, S. (2019). Social and environmental reports at universities: A Habermasian view on their evolution. *Accounting Forum, 43*(3), 283–326.

O'Connor, M., & Spangenberg, J. H. (2008). A methodology for CSR reporting: Assuring a representative diversity of indicators across stakeholders, scales, sites and performance issues. *Journal of Cleaner Production, 16*(13), 1399–1415.

O'Donovan, G. (2002). Environmental disclosures in the annual report. *Accounting, Auditing & Accountability Journal, 15*(3), 344–371.

Owen, D. L., Swift, T., & Hunt, K. (2001, September). Questioning the role of stakeholder engagement in social and ethical accounting, auditing and reporting. *Accounting Forum, 25*(3), 264–282.

Patten, D. M. (2002). The relation between environmental performance and environmental disclosure: A research note. *Accounting, Organizations and Society, 27*(8), 763–773.

Plumlee, M., Brown, D., Hayes, R. M., & Marshall, R. S. (2015). Voluntary environmental disclosure quality and firm value: Further evidence. *Journal of Accounting and Public Policy, 34*(4), 336–361.

Puroila, J., & Mäkelä, H. (2019). Matter of opinion: Exploring the socio-political nature of materiality disclosures in sustainability reporting. *Accounting, Auditing & Accountability Journal, 32*(4), 1043–1072.

Qiu, Y., Shaukat, A., & Tharyan, R. (2016). Environmental and social disclosures: Link with corporate financial performance. *The British Accounting Review, 48*(1), 102–116.

Rezaee, Z., & Tuo, L. (2019). Are the quantity and quality of sustainability disclosures associated with the innate and discretionary earnings quality? *Journal of Business Ethics, 155*(3), 763–786.

Rimmel, G., & Jonäll, K. (2013). Biodiversity reporting in Sweden: Corporate disclosure and preparers' views. *Accounting, Auditing & Accountability Journal, 26*(5), 746–778.

Roberts, R. W. (1992). Determinants of corporate social responsibility disclosure: An application of stakeholder theory. *Accounting, Organizations and Society, 17*(6), 595–612.

Rodrigue, M., Magnan, M., & Cho, C. H. (2013). Is environmental governance substantive or symbolic? An empirical investigation. *Journal of Business Ethics, 114*, 107–129.

Ruiz-Lozano, M., De Vicente-Lama, M., Tirado-Valencia, P., & Cordobés-Madueño, M. (2022). The disclosure of the materiality process in sustainability reporting by Spanish state-owned enterprises. *Accounting, Auditing & Accountability Journal, 35*(2), 385–412.

Sivarajah, U., Kamal, M. M., Irani, Z., & Weerakkody, V. (2017). Critical analysis of Big Data challenges and analytical methods. *Journal of Business Research, 70*, 263–286.

Steenkamp, N. (2018). Top ten South African companies' disclosure of materiality determination process and material issues in integrated reports. *Journal of Intellectual Capital, 19*(2), 230–247.

Stolowy, H., & Paugam, L. (2018). The expansion of non-financial reporting: An exploratory study. *Accounting and Business Research, 48*(5), 525–548.

Suchman, M. C. (1995). Managing legitimacy: Strategic and institutional approaches. *Academy of Management Review, 20*(3), 571–610.

Tate, W. L., Ellram, L. M., & Kirchoff, J. F. (2010). Corporate social responsibility reports: A thematic analysis related to supply chain management. *Journal of Supply Chain Management, 46*(1), 19–44.

Thorne, L., S. Mahoney, L., & Manetti, G. (2014). Motivations for issuing standalone CSR reports: A survey of Canadian firms. *Accounting, Auditing & Accountability Journal, 27*(4), 686–714.

Venturelli, A., Caputo, F., Cosma, S., Leopizzi, R., & Pizzi, S. (2017). Directive 2014/95/EU: Are Italian companies already compliant? *Sustainability, 9*(8), 1385.

Weerathunga, P. R., Xiaofang, C., Nurunnabi, M., Kulathunga, K. M. M. C. B., & Swarnapali, R. M. N. C. (2020). Do the IFRS promote corporate social responsibility reporting? Evidence from IFRS convergence in India. *Journal of International Accounting, Auditing and Taxation, 40*, 100336.

Windsor, D. (2017). Stakeholder responsibilities: Lessons for managers. *In Unfolding Stakeholder Thinking* (pp. 137–153). Routledge.

CHAPTER 5

Materiality in Sustainability Reporting

Abstract This chapters focuses on materiality in sustainability reporting. After an introduction to the concept of materiality in accounting, the chapter explores how materiality has become a fundamental concept in the context of sustainability reporting. Subsequently, the chapter synthesizes the academic debate on sustainability materiality and analyzes how materiality is defined and approached in the main sustainability reporting frameworks and standards. Finally, using a global sample of listed companies, an empirical analysis of the quality of materiality disclosure in sustainability reporting is presented.

Keywords Materiality · Sustainability reporting · Content analysis · Corporate transparency

5.1 THE CONCEPT OF MATERIALITY

Materiality has long been regarded as a cornerstone concept in both the theory and practice of accounting and auditing (Brennan & Gray, 2005; Holstrum & Messier, 1982; Messier et al., 2005). The materiality concept is broadly understood as being concerned with determining what corporate events and issues are considered important enough to be included within reporting, and relatedly, what items are not considered

© The Author(s), under exclusive license to Springer Nature Switzerland AG 2024
C. Mio et al., *Sustainability Reporting*, Palgrave Studies in Impact Finance, https://doi.org/10.1007/978-3-031-58449-7_5

important enough to be included. In separating and distinctly filtering out less important information, materiality enhances reporting clarity, understandability, and integrity.

The term material in accounting first appeared in accounting and auditing texts at the turn of the twentieth century and in US guidance, issued by the American Institute of Certified Public Accountants (AICPA). In contrast, early references to material facts have been traced back to UK legal cases in the 1860s and the Companies Act 1985 (Edgley, 2014).

The importance of this issue is summarized in the following statement from the Discussion Memorandum published by the Financial Accounting Standards Board (FASB) in 1975:

> The concept of materiality pervades the financial accounting and reporting process. It influences decisions regarding the collection, classification, measurement, and summarization of the data concerning the results of an enterprise's economic activities. It also bears on decisions concerning the presentation of that data and the related disclosures in the financial statements. (FASB, 1975, p. 3)

Bernstein (1967, p. 87) defines the concept of materiality as "part of the wisdom of life" since its basic meaning is that there is no need to be concerned with what is not important or with what does not matter. In accounting, the concept of materiality assumes special significance for two basic reasons (Bernstein, 1967). First, most users of accounting information do not comprehend it easily. Consequently, redundancy and the presentation of insignificant data can be misleading and make the task of analysis even more difficult. Second, the concept of materiality is important throughout the auditing process, especially in planning the scope of the audit and conducting the final evaluation of the overall sufficiency of evidential support of assurance regarding the fairness of financial statements. For both planning and evaluation purposes, the precision of audit tests needs to be related to financial statement materiality (Holstrum & Messier, 1982, p. 46).

Not surprisingly, the issue of materiality has attracted academic interest, and accounting and auditing scholars have deeply explored the nature and operationalization of materiality. Detailed reviews of the literature have been conducted by, amongst others, Holstrum and Messier (1982), Brennan and Gray (2005), and Messier et al. (2005). Holstrum and

Messier (1982) categorized the materiality literature before 1982 into four research areas: (1) the nature of the item, (2) the structural form of the decision model, (3) the relative importance of factors used to determine materiality, and (4) materiality thresholds. Messier et al. (2005) review and discuss materiality research conducted after Holstrum and Messier (1982) under two research methods: archival studies—that use audit firm manuals, data from auditor working papers, or published financial statement data, and auditor reports examining materiality decisions—and experimental studies—that examine materiality judgments of users, auditors, and others (judges/ lawyers).

Extant materiality literature has also brought attention to the plethora of ambiguous and contested conceptions of materiality and authors argue that there is a lack of a shared understanding within and outside of the accounting profession (Bolt & Tregidga, 2023; Edgley, 2014; Roberts & Dwyer, 1998). The existence of multiple definitions, along with a lack of detailed guidance, makes materiality "malleable" (Edgley, 2014, p. 255). Edgley (2014) undertakes a genealogical analysis to investigate the historical origins and the conditions that have shaped our knowledge of materiality, finding that different, at times conflicting, expertise and metaphorical discourses have constituted materiality as multiple categories of knowledge objects, including a moral responsibility; a solution to the problem of over-auditing; a solid epistemic foundation for financial reporting; a scientific technique; a quantitative rule of thumb; a risk management concept; and a mysterious shield.

To address the demands for clarity, definitions of materiality and related interpretative guidance have been promulgated by different standard setters and regulators for use by companies and auditors. Table 5.1 reports the most important international definitions of materiality for financial reporting.

While the materiality concept has evolved, and a common understanding is lacking according to Edgley (2014), there is a consensus among various definitions formulated by professional accounting bodies, standard setters boards, and accounting scholars (for a systematic review of the evolution of the definitions of materiality see Chong, 2015). These definitions generally agree on the following features of accounting materiality:

Table 5.1 Definitions of materiality for financial reporting

Standard-setter	Source	Definitions of materiality
Financial Accounting Standards Board (FASB)	Statement of Financial Accounting Concepts No. 8 Conceptual Framework for Financial Reporting Chapter 3, Qualitative Characteristics of Useful Financial Information As Amended in August 2018	"Materiality is entity specific. The omission or misstatement of an item in a financial report is material if, in light of surrounding circumstances, the magnitude of the item is such that it is probable that the judgment of a reasonable person relying upon the report would have been changed or influenced by the inclusion or correction of the item …. No general standards of materiality could be formulated to take into account all the considerations that enter into judgments made by an experienced, reasonable provider of financial information. That is because materiality judgments can properly be made only by those that understand the reporting entity's pertinent facts and circumstances"
Securities and Exchange Commission (SEC)	SEC Staff Accounting Bulletin: No. 99—Materiality August 1999	"Materiality concerns the significance of an item to users of a registrant's financial statements. A matter is "material" if there is a substantial likelihood that a reasonable person would consider it important"
International Accounting Standards Board (IASB)	Conceptual framework for Financial Reporting. As revised in March 2018	"Information is material if omitting, misstating or obscuring it could reasonably be expected to influence decisions that the primary users of general-purpose financial reports (see paragraph 1.5) make on the basis of those reports, which provide financial information about a specific reporting entity. In other words, materiality is an entity-specific aspect of relevance based on the nature or magnitude, or both, of the items to which the information relates in the context of an individual entity's financial report. Consequently, the Board cannot specify a uniform quantitative threshold for materiality or predetermine what could be material in a particular situation"
International Auditing and Assurance Standards Board (IAASB)	International Standard on Auditing 320 Materiality in Planning and Performing and Audit	"Although financial reporting frameworks may discuss materiality in different terms, they generally explain that: Misstatements, including omissions, are considered to be material if they, individually or in the aggregate, could reasonably be expected to influence the economic decisions of users taken on the basis of the financial statements; Judgments about materiality are made in light of surrounding circumstances, and are affected by the size or nature of a misstatement, or a combination of both; and judgments about matters that are material to users of the financial statements are based on a consideration of the common financial information needs of users as a group"

1. *User perspective.* Definitions of accounting materiality are based on the "user utility theory", according to which an item is material for reporting purposes if its magnitude is such that it can influence the decision-making process and the judgment of the intended general users of the report (Messier et al., 2005). In other terms, the materiality of data and items for reporting purposes is conceived from a user perspective, and it is related to the minimum amount of omission or misstatement that would influence economic decisions made by users of the information (Brennan & Gray, 2005). Therefore, the determination of materiality is based on a consideration of the common financial information needs of users as a group. Concerning the issue of "material to whom", the IASB refers "to primary users of general-purpose financial reports" (i.e., investors) while the FASB and the SEC focus on the perspective and decision-making of a "reasonable person". Holstrum and Messier (1982, 48) pointed out three main problems with a user approach to materiality. First, little is known about how financial statement information is utilized by users in investment and credit decision-making. Second, materiality decisions are made by preparers, auditors, and users; these heterogeneous groups are likely to have dissimilar views of materiality because of their different incentives. Third, there is little information on how materiality judgments made by preparers and auditors affect users' decisions.

2. *Principles-based.* Definitions of accounting materiality are general and principles-based since they do not provide specific numerical-based guidance or thresholds on how to determine if an item is material (Eccles et al., 2012).

3. *Entity-specific.* The determination of materiality is entity-specific, which is consistent with defining materiality in the context of all the other information in an entity's financial report. In the determination of materiality, preparers are required to take into consideration all the surrounding (quantitative and qualitative) facts and circumstances, such as "the nature or magnitude, or both, of the items to which the information relates in the context of an individual entity's financial report" (IASB, 2018) and "the reporting entity's pertinent facts and circumstances" (FASB, 2018).

4. *Professional judgements.* The materiality assessment process is a matter of professional judgment and no set of rules can be applied to determine thresholds in all circumstances. Decisions about materiality take into consideration the surrounding circumstances, and the examination of what is important to users remains, to a certain extent, a highly personal evaluation of the decision-makers. Bernstein (1967) observes that in few areas of accounting practice is the term "judgment" invoked so categorically as in the area of materiality determination, so that in a profession where objectivity is a consideration of cardinal importance, "materiality seems to be its "Achilles' Heel" (p. 90).

5.2   Sustainability Materiality Research

While historically, materiality has typically been viewed within financial reporting, more recently the materiality concept has also become a core focus in sustainability reporting (Unerman & Zappettini, 2014; Puroila & Mäkelä, 2019). Applying a materiality filter for sustainability reporting requires companies to identify, assess, and prioritize the most relevant social, environmental, and economic sustainability information to be communicated and included in a sustainability report. Practically speaking, the goal is to distinguish between material sustainability issues, i.e., those sustainability issues that are likely to influence the decision-making of stakeholders, and those issues that are non-material, i.e., not likely to influence stakeholder decision-making (Jørgensen et al., 2022, p. 344).

The aim of these processes appears to be similar to the aim of materiality in financial reporting.

Sustainability materiality is intended to increase transparency and accountability by making the sustainability reports more focused on "what matters" to companies and stakeholders, and ensure that the clarity of the report is not diminished by excessive detail on relatively minor issues. This ensures that sustainability reporting adheres to high standards of quality and effectively communicates relevant information about the company.

The imperative of a materiality hurdle or filter to select and focus on the most relevant issues for communication about the company, while omitting minor issues in a report, is even more pressing for ensuring clarity in sustainability reporting compared to financial reporting. Indeed, due to the vastness and heterogeneity of sustainability information available, the problem of information overload in a sustainability report is even greater than the scope for information overload in a financial report (Baumüller & Schaffhauser-Linzatti, 2018; Unerman & Zappettini, 2014, p. 175). By providing "legitimate closure" to the reporting content, and the variety of understandings of corporate sustainability, the assessment of materiality is understood more as a socioeconomic and political phenomenon than a technical one (Puroila & Mäkelä, 2019). Furthermore, enhancing the focus on materiality in sustainability reporting is considered an important remedy against the threat of greenwashing (Baumüller & Schaffhauser-Linzatti, 2018, p. 102).

Notwithstanding certain shared characteristics, the meaning of materiality is not the same in both financial and sustainability reporting contexts. While the purpose of discerning between more or less material issues in financial reporting is related to understanding factors influencing the financial performance of the company, materiality in sustainability reporting is more related to the company's social and environmental performance (Jørgensen et al., 2022, p. 344). Therefore, in the context of sustainability reporting, the assessment of materiality is more complex because it involves taking into consideration the multiple, and often divergent, perspectives of different stakeholder groups. Puroila and Mäkelä (2019) argue that the introduction of materiality within sustainability reporting has the potential to challenge the narrow scope and the dominance of shareholder interests within materiality assessments. In other terms, when extended into the context of sustainability reporting, materiality has shifted toward a new stakeholder logic (as opposed to a purely market logic) emphasizing the social and environmental impacts of corporate non-financial performance and the importance of stakeholder engagement (Edgley et al., 2015). Introducing the concept of materiality into the arena of sustainability reporting requires companies to identify what matters taking into consideration a broader set of stakeholders with multiple informational needs, instead of focusing merely on financial considerations and the interest of investors. In effect, the main objective of the materiality principle is to provide stakeholders with information consistent with their needs and expectations, so that they can assess the company's performance.

Furthermore, and related to this, a materiality filter appears even more difficult to apply in the context of sustainability reporting than in financial settings (Lai et al., 2017; Mio et al., 2019; Reimsbach et al., 2020).

As argued by Lai et al. (2017, p. 535) sustainability information seeks to capture a broader concept of value creation and not only this information sometimes is not quantifiable, but even when the impact of an event may be quantified, establishing a unique quantitative threshold to define the materiality of such events is not always feasible. Furthermore, the materiality of sustainability-related topics has to be evaluated considering the informative needs of different stakeholder categories (Deegan & Rankin, 1997), which may have divergent demands and expectations (Reimsbach et al., 2020).

Applying materiality hurdles raises questions about who decides whether a particular issue is relatively major or minor to be or not to be communicated or from whose perspective an issue is material or immaterial. Materiality, when viewed as a relative judgment of value, inherently requires a process encompassing the selection, inclusion, and exclusion of information (Unerman & Zappettini, 2014, p. 177). To address these issues, the main sustainability reporting standards and frameworks now set out materiality processes to be adopted that can guide reporting organizations in deciding whether a particular sustainability item is or is not sufficiently material to be disclosed in the sustainability report (see par. 3.3 for an overview of materiality in sustainability reporting frameworks).

Not surprisingly, the notion of materiality in the context of sustainability reporting has attracted growing academic interest. Fiandrino et al. (2022) undertake a systematic review of the extant literature related to sustainability materiality. The findings of their review underscore a significant surge in scholarly works about sustainability materiality since the onset of the 2010s. Furthermore, Fiandrino et al. (2022) identify eight distinct thematic areas in the broader stream of sustainability materiality research:

1. *Definitions of Materiality.* Academics discuss the landscape of materiality definitions provided by sustainability reporting standards (Christensen et al., 2021; Clark, 2021; Jørgensen et al., 2022) and regulation (Baumüller & Schaffhauser-Linzatti, 2018; La Torre et al., 2020) as well as how materiality is understood by preparers (Lai et al., 2017) and participants from the non-financial reporting context (Bolt & Tregidga, 2023). These studies highlight that the various definitions of sustainability materiality are not independent of the purpose of reporting and the users of sustainability information (Fiandrino et al., 2022). For instance, Lai et al. (2017) find that

the meaning that integrated reporting preparers attach to materiality corresponds with the company strategy and that although several actors engage in integrated reporting preparation, the materiality determination process is governed by a specific "IR hub" in strict collaboration with and dependence on the chief financial officer. Bolt and Tregidga (2023) demonstrate that participants who engage with the materiality concept in different ways (e.g., academics, auditors, assurors, consultants, and standard setters) tend to use stories and narratives, particularly stories of materiality "in action", to navigate materiality complexity and ambiguity and make and give sense to the concept. This reveals a disconnect between the static, technical definitions of materiality currently favored by standard setters and guidance providers, and the creative authoring processes the participants employ.

2. *Pressures over materiality analysis.* Academics have examined some pressures and concerns related to sustainability reporting—such as greenwashing behavior, rhetorical or symbolic representation of sustainability, ceremonial reporting processes, moral fictionalism, and different methodologies for materiality applications—that may impede to conduct and disclose of proper sustainability materiality analysis. Using interviews with 20 preparers from 14 large organizations listed on the Johannesburg Stock Exchange, Cerbone and Maroun (2020) find that market-dominated firms have an internally focused approach to setting materiality which emphasizes value relevance for financial capital providers. In that case, companies with a finance-centric market logic fail to provide a comprehensive account of the value creation process and how the business ensures long-term sustainability.

3. *Materiality determinants and indicators.* Based on their systematic literature review, Fiandrino et al. (2022) find that existing evidence suggests that sustainability materiality is positively associated with proactive sustainable behaviors and that prior academic studies have identified the following determinants of materiality: learning effects, gender diversity, the assurance of sustainability information, involvement of board members, general factors related to company size, industry, country, and the complexity of companies.

4. *Issues that are material for companies and stakeholders.* Prior studies have explored the materiality of sustainability issues for companies and certain categories of stakeholder groups. Eccles et al.

(2012) investigated climate change-related issues disclosed in a 10-K filing submitted to the Securities and Exchange Commission in 2011 by 48 companies of six different industries and noted the lack of material information disclosed in a comparable format, due to differences among industries. Based on a sample of 10 large companies adopting the G4 guidelines, Jones et al. (2016) find marked variations in the ways, and the extent to which, companies apply materiality filters and that many of the high-priority material issues identified by these companies are centered on business continuity rather than environmental sustainability issues. Whitehead (2017) finds that for the New Zealand wine industry, environmental issues make up the highest priority issues, followed by social issues relating primarily to worker wellbeing. Reimsbach et al. (2020) adopt a user perspective and show how and why NFI affects stakeholder decisions and how the different decision-making processes lead to a state where the materiality of NFI ultimately lies in the eye of the beholder. The results of their experimental study reveal that employees evaluate NFI as more material than investors, but both energy and biodiversity have equal importance. The user perspective is also adopted by Deegan and Rankin (1997) who investigate, by way of a questionnaire, the materiality of environmental issues to certain groups in society who use annual reports to gain information. Findings indicate that shareholders and individuals within organizations with a review or oversight function consider that environmental information is material to the particular decisions they undertake, while stockbrokers and analysts downplay the materiality of environmental information.

5. *The evaluation of materiality in sustainability information.* A significant amount of research has explored the quality of materiality disclosure in sustainability reports or integrated reports (Beske et al., 2020; Machado et al., 2021; Ruiz-Lozano et al., 2022) and its determinants (Fasan & Mio, 2017; Farooq et al., 2021; Gerwanski et al., 2019; Torelli et al., 2019). Extant evidence suggests that companies disclose only a small amount of information related to the materiality assessment process (Machado et al., 2021; Ruiz-Lozano et al., 2022) and they fail to explain the methods for stakeholder engagement and topic identification (Beske et al., 2020). About the determinants of sustainability materiality disclosure, Fasan and Mio (2017) find that industry and some firm-level characteristics

(i.e., board size and diversity) play a significant role in driving materiality disclosure in firms' integrated reports. Torelli et al. (2019) provide empirical evidence that the degree of breadth and depth of implementation of the materiality analysis process for sustainability reporting purposes is positively related to stakeholder engagement and the implementation of the GRI guidelines. Finally, Farooq et al. (2021) document that materiality assessment disclosure is positively influenced by higher financial performance, lower leverage, and better corporate governance. Other determinants of materiality disclosure quality include learning effects, gender diversity, and assurance (Gerwanski et al., 2019). Puroila and Mäkelä (2019) conducted a qualitative content analysis of materiality disclosures in 44 sustainability reports from companies listed in the Global 100 Index that follow the GRI guidelines. They find that, despite the seemingly rigorous, objective, and technical exercise of defining a set of material corporate sustainability issues, the assessment of materiality and the resulting presentation of corporate sustainability is a value-laden, political judgment of what matters in corporate sustainability, favoring the corporate financial interests and falling short of addressing the complexity of sustainable development.

6. *Models of materiality assessment.* Given the relevance of stakeholder engagement in the materiality determination process, academics have proposed several models to support companies' materiality assessment of sustainability aspects and indicators from the viewpoint of the targeted stakeholder groups (Calabrese et al., 2015; Hsu et al., 2013). For instance, Calabrese et al. (2015) propose a CSR model that provides a simultaneous and systematic assessment of three important aspects of CSR commitment (company-disclosed, customer-perceived, and customer-required). This model can support companies in assessing the materiality of CSR aspects and indicators from the viewpoint of the targeted stakeholder groups (e.g., customers), through their classification based on their CSR feedback. Hsu et al. (2013) constructed a model of materiality analysis for determining material issues to be included in sustainability reporting in accordance with stakeholders needs based on failure modes and effects analysis (FMEA), that is used to establish the evaluation criteria and construct a model of materiality analysis of risk priority numbers (RPNs), and analytic network

process (ANP) method, that is used to determine the related weight of the criteria.

7. *Impact of material information and value relevance of materiality.* Academics debate the impacts of materiality on the financial performance of firms (Carvajal & Nadeem, 2023; Consolandi et al., 2022; Khan et al., 2016; Schiehll & Kolahgar, 2021). The seminal paper of Khan et al. (2016) has brought forward evidence of the importance of ESG materiality in investment decision-making. Adopting the concept of materiality of the Sustainability Accounting Standards Board (SASB) and working on a sample of more than 2,000 US companies over 21 years, Khan et al. (2016) demonstrate that companies top performing on ESG material issues outperform those in the bottom quantile and that high performance on immaterial ESG issues doesn't lead to superior financial performance. Based on a large sample of US companies from January 2008 to July 2019, Consolandi et al. (2022) find that the market seems to reward those companies operating in industries with a high level of concentration of ESG materiality. Consequently, not only does the performance on ESG material issues have a consistent impact on stock performance, but also that what matters more is the level of concentration of materiality: having fewer material issues but more financially relevant ones is rewarded by the market. Furthermore, Schiehll and Kolahgar (2021) conducted an automated content analysis of 150,000 electronic documents filed by firms listed on the S&P/TSX Composite Index from 1999 to the end of 2014, to find not only that ESG disclosure is value relevant but also that financial materiality at ESG disclosure provides incremental effects on a firm's stock price informativeness compared with overall ESG disclosure. Thus, stock price informativeness is greater when a firm's ESG disclosure is measured according to the SASB-identified material information framework. Similarly, Carvajal and Nadeem (2023) find that, in the context of New Zealand, firm performance has a positive relationship with both disclosure of sustainability information and financial material sustainability information. The findings indicate that disclosure of financially material sustainability information has a stronger relationship with firm performance, showing a significant association with both ROA and ROE, compared with the disclosure of sustainability information itself, which has a positive relationship with ROA only.

8. *Materiality in sustainability assurance*. Scholars explore materiality in the context of sustainability reporting assurance and investigate how sustainability materiality assessments differ from financial statement materiality assessments (Canning et al., 2019; Edgley et al., 2015; Moroney & Trotman, 2016). Edgley et al. (2015) show that materiality in social and environmental reporting is significantly different from materiality in financial audit: sustainability materiality is characterized by a new stakeholder logic (as opposed to a purely, short term, market logic), which requires auditors to recognize the value of stakeholder engagement in identifying materials issues. Furthermore, sustainability materiality not only considers past data but is a forward-looking lens and it has a more subjective nature than materiality in financial audit. Moroney and Trotman (2016) find that auditors assess the materiality of an audit difference significantly higher for a financial case than for a sustainability case and that qualitative factors have a greater impact on sustainability materiality assessments than on financial statement materiality assessments when an audit difference is between 5% and 10% of a relevant base. In their examination of how materiality has been translated from financial audit to the sustainability assurance arena, Canning et al. (2019) illustrate how assuror flexibility, underpinned by assuror intuition, is central to uncovering assurance technologies deemed capable of addressing the materiality of ambiguous sustainability data and that assurors with no financial audit background retrospectively rationalize their intuition using the assumed authority of structured financial audit methodologies. This facilitates the tentative translation of financial audit knowledge to the sustainability assurance domain.

5.3   THE TYPES OF MATERIALITY
IN SUSTAINABILITY REPORTING STANDARDS

This paragraph examines the plurality of definitions that the concept of materiality has in the sustainability reporting standards. Since a particular piece of information is considered "material" if it could influence the decision-making of users concerning the reporting company, materiality is not a clear-cut concept and is open to different approaches and interpretations. Therefore, firms face a large amount of discretion when it comes to self-assessing the materiality of their sustainability issues, leading to inherent subjectivity in the materiality assessment. To assist firms in their materiality analysis, and ensure high-quality and consistent sustainability reporting, many sustainability reporting standards and frameworks—such as the GRI, Sustainability Accounting Standards Board (SASB), International Integrated Reporting Council (IIRC), Task Force on Climate-Related Financial Disclosures (TCFD), the European Sustainability Reporting Standards (ESRS) and International Sustainability Reporting Standards (ISRS)—define the materiality concept and set out guidance to assess whether a particular sustainability issue is sufficiently material to be disclosed or not in a sustainability report (Clark, 2021; Jørgensen et al., 2022).

Table 5.2 presents a synthesis of the definitions of materiality provided by the main sustainability reporting standards and frameworks (for similar syntheses see also Abhayawansa, 2022; Cooper & Michelon, 2022; Fiandrino et al., 2022).

The concept of materiality adopted in sustainability reporting has been closely scrutinized in the literature (Clark, 2021; Cooper & Michelon, 2022), in particular in the context of the GRI standards (Perera-Aldama, 2023), that have also been contrasted with other frameworks (De Villiers et al., 2022; Jørgensen et al., 2022; Pizzi et al., 2022), and in the recent developments at the level of the EU regulation (Baumüller & Schaffhauser-Linzatti, 2018; Baumüller & Sopp, 2022; La Torre et al., 2020) and the new global standards issued by the ISSB (Abhayawansa, 2022; Giner & Luque-Vílchez, 2022; Stolowy & Paugam, 2023).

Scholars point out that both similarities and differences exist in the definitions of materiality in sustainability reporting standards. On the one hand, the main sustainability standard-setting bodies converge on a set of common foundational principles on materiality (Corporate Reporting Dialogue, 2016). In particular, they all explicitly rely on a user perspective, where the materiality of information is judged by its potential effect on the users' decisions. Moreover, Baumüller and Schaffhauser-Linzatti (2018) note that in all cases, the determination of materiality is considered to be "a complex multi-step process that requires judgments which take into consideration both qualitative and quantitative aspects, although these aspects are differently weighted. Decision usefulness is a key criterion in all frameworks, but viewed from different perspectives. Furthermore, materiality is highly specific to the individual context

Table 5.2 Definitions of materiality in sustainability reporting standards/frameworks

Standard setter	Standard/framework	Definition of materiality	Type of materiality	Intended audience
AccountAbility	AA1000 Accountability Principles (2018)	"Materiality relates to identifying and prioritising the most relevant sustainability topics, taking into account the effect each topic has on an organisation and its stakeholders. A material topic is a topic that will substantively influence and impact the assessments, decisions, actions and performance of an organisation and/or its stakeholders in the short, medium, and/or long term" (p. 20)		Organization and stakeholders
European Commission	Directive 2014/95/EU	The management report shall include "a non-financial statement containing information to the extent necessary for an understanding of the undertaking's development, performance, position and impact of its activity, relating to, as a minimum, environmental, social and employee matters, respect for human rights, anti-corruption and bribery matters"	Double materiality	Multiple stakeholders
European Commission	Directive EU 2022/2464	The management report shall include "information necessary to understand the undertaking's impacts on sustainability matters, and information necessary to understand how sustainability matters affect the undertaking's development, performance and position"	Double materiality	Multiple stakeholders
European Sustainability Reporting Standards (ESRS)	ESRS 1 General Requirements (2023)	"The undertaking shall report on sustainability matters based on the double materiality principle" (par. 21). "Double materiality has two dimensions, namely: impact materiality and financial materiality" (par. 37)	Double materiality	Multiple stakeholders
Global Reporting Initiative (GRI)	GRI 1: Foundation 2021	"When using the GRI Standards, the organization prioritizes reporting on those topics that represent its most significant impacts on the economy, environment, and people, including impacts on their human rights. In the GRI Standards, these are the organization's material topics" (p. 8)	Impact materiality	Multiple stakeholders

(continued)

Table 5.2 (continued)

Standard setter	Standard/ framework	Definition of materiality	Type of materiality	Intended audience
IFRS Sustainability Disclosure Standard	IFRS S1 General Requirements for Disclosure of Sustainability-related Financial Information (2023)	"An entity shall disclose material information about the sustainability-related risks and opportunities that could reasonably be expected to affect the entity's prospects. In the context of sustainability-related financial disclosures, information is material if omitting, misstating or obscuring that information could reasonably be expected to influence decisions that primary users of general purpose financial reports make on the basis of those reports, which include financial statements and sustainability-related financial disclosures and which provide information about a specific reporting entity" (par. 17–18)	Financial materiality	Existing and potential investors, lenders, and other creditors
International Integrated Reporting Council (IIRC)	International, <IR> Framework (2021)	"An integrated report should disclose information about matters that substantively affect the organization's ability to create value over the short, medium, and long term" (par. 3.17)	Financial materiality	Investors and providers of financial capital
Sustainability Accounting Standards Board (SASB)	SASB Conceptual Framework (2017)	"SASB standards address the sustainability topics that are reasonably likely to have material impacts on the financial condition or operating performance of companies in an industry…. In identifying sustainability topics that are reasonably likely to have material impacts, the SASB applies the definition of "materiality" established under the US securities laws. According to the US Supreme Court, information is material if there is "a substantial likelihood that the disclosure of the omitted fact would have been viewed by the reasonable investor as having significantly altered the 'total mix' of information made available" (p. 9)	Financial materiality	Investors

Standard setter	Standard/ framework	Definition of materiality	Type of materiality	Intended audience
Task Force on Climate-Related Financial Disclosures (TCFD)	Recommendations of the Task Force on Climate-related Financial Disclosures (2017)	"Better disclosure of the financial impacts of climate-related risks and opportunities on an organization is a key goal of the Task Force's work. In order to make more informed financial decisions, investors, lenders, and insurance underwriters need to understand how climate-related risks and opportunities are likely to impact an organization's future financial position as reflected in its income statement, cash flow statement, and balance sheet (p. 8)	Financial materiality	Investors, lenders, and insurance underwriters
World Economic Forum	Measuring Stakeholder Capitalism Toward Common Metrics and Consistent Reporting of Sustainable Value Creation (2020)	"This project uses the term "material" to mean information that is important, relevant and/or critical to long-term value creation. … Our perspective is that the recommended metrics reflect not only financial impacts but "pre-financial" information that may not be strictly material in the short term, but are material to society and planet and therefore may become material to financial performance over the medium or longer term. Materiality is a dynamic concept, in which issues once considered relevant only to social value can rapidly become financially material" (p. 14)	Dynamic materiality	Multiple stakeholders

and thus cannot be appropriately addressed by simple checklist-based approaches" (p. 106).

Nevertheless, different understandings, definitions, and approaches have emerged over the years since various sustainability reporting frameworks and standards are based on different objectives and allow firms to develop a system of disclosure tailored to the unique needs of specific stakeholder groups. The operationalization of the definitions of materiality varies in three key aspects, as identified by Clark (2021). First is the intended audience of the report, which is the type of stakeholder whose sustainability information needs are to be served (i.e., Are these only financial decision-makers such as investors and financiers? Or do they also include other parties such as employees, suppliers, customers, and communities; that is, the socioeconomic environment?). The second is the content of the information, while the third refers to the time horizon considered by the intended users of the information.

Taking into account these differences, materiality definitions adopted by sustainability reporting standards and frameworks can be categorized into four main groups: *impact* materiality, *financial* materiality, *double* materiality, and *dynamic* materiality.

As noted by Abhayawansa (2022) differences exist between the definitions even within the same group and the definitions provide much scope for interpretation, requiring significant professional judgment for operationalizing them. Because of the malleable nature of the materiality definitions, materiality is said to remain an ambiguous (Puroila & Mäkelä, 2019) and contested (Reimsbach et al., 2020) concept, which can be adapted to suit different perspectives about the purpose of sustainability reporting (Cooper & Michelon, 2022). Abhayawansa (2022) considers the principle of materiality to be the "most divisive concept in the regulatory work on standard setting for sustainability reporting" (p. 1363).

The principles of impact and financial materiality represent the foundational and traditional approaches that form the basis for the newly emerged concepts of double materiality and dynamic materiality. The divergence between impact and financial materiality has been traditionally best illustrated by the different approaches adopted in the Sustainability Accounting Standard Boards (SABS) and the Global Reporting Initiative (GRI & SASB, 2021; Jørgensen et al., 2022; Reimsbach et al., 2020), which represent the opposite sides of a spectrum ranging from the consideration of a single stakeholder group, i.e., investors (SASB), to the consideration of all possible stakeholders (GRI), and reflects the tensions between understanding sustainability reporting as a tool for financial decision-makers or a means to engender sustainable development (Abhayawansa, 2022).

Building on a sample of 2,046 US-listed companies observed during the period 2017–2020, Pizzi et al. (2022) finds that financial and impact materiality are driven by different purposes: while the SASB's adoption (financial materiality) is driven by factors related to financial dynamics, the GRI's adoption (impact materiality) is influenced by the existence of corporate governance mechanisms inspired by sustainable and ethical principles.

The notion of materiality continues to be a divisive concept also considering the recent developments in the sustainability reporting field, which in 2020 witnessed the entry of two institutions: the European Commission in collaboration with the European Financial Reporting Advisory Group (EFRAG) and the International Financial Reporting Standards (IFRS) Foundation, with which the newly established International Sustainability Standard Board (ISSB) is charged with and which consolidated the Value Reporting Foundation (formed in June 2021 as the combination of the IIRC and the SASB), the TCFD and the Climate Disclosure Standards Board. The notion of materiality is among the most important aspects that differ between the approaches of the European Commission/EFRAG and the ISSB (De Villiers et al., 2022; Giner & Luque-Vílchez, 2022). On the one hand, the sustainability reporting standards issued by the ISSB have an exclusive focus on financial materiality for an investor-based audience, while the European Commission adopts a double materiality approach for the benefit of multiple stakeholders.

Significant tensions between the different approaches to materiality emerged when they were applied in practice, given that it is challenging to discern between financially material sustainability issues and those that are (only) material from the impact perspective. Adopting different approaches to materiality is all the more problematic because they can create confusion and may lead users of sustainability reports to draw unjustified conclusions based on materiality assessments (Jørgensen et al., 2022). Thus, users must receive clear claims in a manner that unequivocally communicates which concept of materiality is adopted in a firm's sustainability reporting (Jørgensen et al., 2022).

As emerged from the study of Jørgensen et al. (2022) the dividing line between these rival approaches to materiality (impact vs. financial materiality) goes between "those that believe that sustainability reporting should first and foremost be a tool for better-informing decision-makers in financial markets about the financial implications of sustainability issues and those that believe that sustainability reporting should serve to increase the transparency of sustainability impacts in a manner that can drive real sustainability improvements" (p. 356). Similarly, Abhayawansa (2022) contends that the recent debate about the definition of materiality in sustainability reporting is mainly underscored by differences in views about the intended audience (i.e., whose needs corporate sustainability reports should serve) (p. 1371).

Each type of sustainability materiality is discussed in the following sections.

Impact Materiality

Impact materiality (or social and environmental materiality) is characterized by an inside-out approach: topics are material if they reflect relevant impacts (i.e., externalities) the reporting organization has on the economy, environment, and/or society. Impact materiality assumes that sustainability reporting plays an accountability role, according to which *all stakeholders* are conceived as the primary addresses and users of such reporting (Cooper & Michelon, 2022).

As emerged from the different definitions of materiality according to GRI over the years discussed by Perera-Aldama (2023), the best example of sustainability reporting standards building on the notion of impact materiality is the GRI standards.

The materiality perspective is strictly related to the objective of sustainability reporting using the GRI Sustainability Reporting Standards, which is "to provide transparency on how an organization contributes or aims to contribute to sustainable development" (GRI, 2021a, p. 7).

The notion of materiality in the context of GRI has been revised with the updated system of GRI Standard in 2021 (Perera-Aldama, 2023). In the *GRI Universal Standards 2016*, material topics were topics that reflect at least one of the following dimensions:

a. the organization's significant economic, environmental, and social impacts;
b. their substantive influence on the assessments and decisions of stakeholders.

So, a distinction was made between two dimensions which both can induce materiality independently: "impacts" and "stakeholders". Combining these two dimensions results in the so-called materiality matrix, as recommended by the *GRI 101: Foundation 2016*. This matrix has "Influence on stakeholder assessments and decisions" along the Y-axis and "Significance of the reporting organization's economic, environmental, and social impacts" along the X-axis.

The focus lies on the intersection of both dimensions in the matrix, however, a topic can also be material based only on one of the two dimensions.

This approach was revised by the GRI because feedback indicated that it "often led to biases and incorrect interpretations of these dimensions. Separating impact assessment from identifying stakeholder views left materiality assessments particularly vulnerable to biases based on stakeholder selection, given that this approach led organizations to prioritize impacts only if the consulted stakeholders highlighted them" (GRI, 2022, p. 17).

The new universal standard *GRI 1: Foundation 2021* defines material topics as topics "that represent the organization's most significant impacts on the economy, environment, and people, including impacts on their human rights" (GRI, 2021a, p. 31). Therefore, the "influence on the assessments and decisions of stakeholders" is no longer a stand-alone factor that determines whether a topic is material. While the definition of "material topic" has been revised to focus on impact, engagement with relevant stakeholders forms part of identifying and assessing an organization's impacts and informs the process for determining material topics.

Furthermore, Universal Standard *GRI 3 Material Topics 2021* presents a mandatory shift requiring organizations to "review each topic described in the applicable Sector Standards and determine whether it is a material topic for the organization" (GRI, 2021b, p. 15).

Guidance to determine material topics is provided by the standard *GRI 3: Material Topics 2021*, which describes the four steps that an organization should follow in determining its material topics:

1. *Understand the organization's context.* The organization creates an initial high-level overview of its activities and business relationships, the sustainability context in which these occur, and an overview of its stakeholders.

2. *Identify actual and potential impacts.* In this step, the organization identifies its actual and potential impacts on the economy, environment, and people, including impacts on human rights, across the organization's activities and business relationships. These impacts include negative and positive impacts, short-term and long-term impacts, intended and unintended impacts, and reversible and irreversible impacts. In this step, the organization needs to consider the impacts described in the applicable GRI Sector Standards and determine whether these impacts apply to it.

3. *Assess the significance of the impacts.* The organization may identify many actual and potential impacts. In this step, the organization assesses the significance of its identified impacts to prioritize them. With regard to negative impacts, the significance of an actual negative impact is determined by its severity, which is determined by its scale (how grave the impact is), scope (how widespread the impact is), and irremediable character (how hard it is to counteract or make good the resulting harm), while the significance of a potential negative impact is determined by the severity and likelihood of the impact. In the case of positive impacts, the significance of an actual positive impact is determined by the scale and scope of the impact while the significance of a potential positive impact is determined by the scale and scope as well as the likelihood of the impact.

4. *Prioritize the most significant impacts for reporting.* In this step, to determine its material topics for reporting, the organization prioritizes its impacts based on their significance. The significance of an impact is assessed in relation to the other impacts the organization has identified. The organization should arrange its impacts from most to least significant and define a cut-off point or threshold to determine which of the impacts it will focus its reporting on.

Financial Materiality

Financial materiality is defined as the "selection of information based on economic value creation at the level of the reporting entity for the benefit of capital providers" (Abhayawansa, 2022, p. 1364). Financial materiality is based on an outside-in approach: sustainability topics are material depending on their ability to affect the reporting organization. Financial materiality is based on the assumption that sustainability reporting plays a valuation or stewardship role (Cooper & Michelon, 2022), both of which conceive financial stakeholders, i.e., shareholders (and potential investors) and creditors, as the main users of sustainability reporting.

Materiality definitions provided by the SASB (2017), IIRC (2021), TCFD (2017), and, more recently, the ISSB (2023), fall within this group.

SASB's approach to materiality is based on a financially-oriented definition that is well-accepted by capital markets globally. The SASB Standards identify sustainability topics that are reasonably likely to impact the financial performance and long-term enterprise value of the typical company in an industry (GRI & SASB, 2021). SASB provides standards for 77 industries across 11 sectors and each standard identifies the subset of sustainability issues reasonably likely to impact the financial performance and long-term enterprise value of the typical company in an industry.

SASB focuses on disclosure by companies to their investors and other providers of financial capital. For a sustainability issue to translate into financial performance, it must have an impact on either the amount of cash flow generated by the company or the cost of external financing to the company. Therefore, any information that only has a non-economic impact on the firm and affects decisions other than those related to investment is, by definition, not material.

Also, the definition of materiality adopted in the sustainability reporting standards issued by the ISSB is related to the concept of financial materiality.

The IFRS S1 *General Requirements for Disclosure of Sustainability-related Financial Information* defines material information in alignment with the definition in IASB's Conceptual Framework for General Purpose Financial Reporting. The ISSB requires an entity to disclose material information about the "sustainability-related risks and opportunities that could reasonably be expected to affect the entity's prospects" (ISSB,

2023, par. 17) and states that information is material for "sustainability-related financial information" if "omitting, misstating or obscuring that information could reasonably be expected to influence decisions that primary users of general purpose financial reports make on the basis of those reports" (ISSB, 2023, par. 18). The assessment of materiality shall be made in the context of the information necessary for users of general-purpose financial reporting (i.e., existing and potential investors, lenders and other creditors) to assess enterprise value and in making decisions relating to providing resources to the entity.

Scholars highlight the drawbacks of an exclusive focus on financial materiality, according to which the identification of material matters is limited to what impact an organization's ability to create value. Abhayawansa (2022, p. 1374) notes that "A narrow view of enterprise value is detrimental to the capital market in the long run (not to mention the planet) and deters organisations from focusing on contributing to sustainable development in line with the United Nations Sustainable Development Goals". Jørgensen et al.'s (2022) survey findings reveal that even investors disagree with the statement that sustainability reporting should account only for financially material sustainability issues.

Double Materiality

Double materiality is defined as the merger between impact and financial materiality. According to this approach, material sustainability information is any information that is material from the impact or the financial perspective. It establishes that sustainability issues and information should be assessed through both an "outside-in" perspective, about their effects on companies' financial performance, and an "inside-out" perspective concerning the impacts on the environment and society. This definition provides a holistic perspective on materiality, which corresponds with the interest of both stakeholders at large and shareholders or investors specifically. Thereby, the double materiality "encloses an ideological conflict between the investors' financial interests and other stakeholders' needs" (De Villiers et al., 2022, p. 737).

Double materiality is the approach adopted by the European Commission in the regulation of mandatory sustainability reporting, first in the Non-Financial Reporting Directive (NFRD) (Directive 2014/94/EU) and, more explicitly, in the Corporate Sustainability Reporting Directive

(Directive EU 2022/2464) (Baumüller & Sopp, 2022; La Torre et al., 2020).

As Baumüller and Schaffhauser-Linzatti (2018) explain, in designing the EU Directive, the EU Commission acknowledged the importance of the materiality of NFI to avoid information overload and greenwashing in non-financial reporting. However, quite surprisingly, neither the terms "materiality" nor "material" are used throughout the reporting requirements of the NFRD. Nevertheless, despite different wordings, several provisions of the NFRD refer to the principle of materiality.

The starting point for any question related to materiality assessment in the context of the NFRD is Art. 19a (1) which provides a definition of information that should be disclosed, by stating that the companies concerned:

> [...] shall include in the management report a non-financial statement containing information to the extent necessary for an understanding of the undertaking's development, performance, position and impact of its activity, relating to, as a minimum, environmental, social and employee matters, respect for human rights, anti-corruption and bribery matters.

With respect to this article, the *Guidelines on Nonfinancial Reporting* (2017/C 215/01) issued by the European Commission in July 2017 formulate the key principle called "Disclose material information" (European Commission, 2017). The NFRD addresses two main types of information that could qualify as material:

1. information necessary for an understanding of the undertaking's development, performance, and position;
2. information necessary for an understanding of the undertaking activities' impacts.

The majority of commentators agree that since those two points are linked with the word "and" in the legislation, both (financial relevance for the company and financial relevance in terms of impact) have to be met simultaneously for information to qualify as material. Then, information that refers to certain impacts but is of no financial relevance is not covered by the reporting obligations under the NFRD (Baumüller & Schaffhauser-Linzatti, 2018, p. 107; Baumüller & Sopp, 2022; La Torre et al., 2020). Based on this interpretation, Baumüller and Schaffhauser-Linzatti (2018) observe that the NFRD establishes a regime for assessing materiality that builds on materiality for financial reporting purposes, where the providers of financial capital receive special attention, but introduces a new element to be taken into account when assessing the materiality of NFI by referring to information "to the extent necessary for an understanding of the [...] impact of (the company's) activity".

The term double materiality was first introduced explicitly in the *Guidelines on Non financial Reporting: Supplement on Reporting Climate-related Information* (2019/C 209/01) issued by the European Commission in June 2019 (Baumüller & Sopp, 2022). These Guidelines include a clarification of the proper interpretation of the principle of materiality for non-financial reporting and explain that the NFRD adopts a double materiality perspective (European Commission, 2019, p. 4), where:

- the reference to the company's "development, performance [and] position" indicates financial materiality, according to which sustainability-related information should be reported if necessary for an understanding of the value of the company (not just in the sense of affecting financial measures recognised in the financial statements) typically for the interest of investors;
- the reference to "impact of [the company's] activities" indicates impact materiality, according to which sustainability-related information should be reported if it is necessary for an understanding of the external impacts of the company, typically for the interest of citizens, consumers, employees, business partners, communities and civil society organisations.

With these guidelines, the European Commission explicitly acknowledges that materiality in sustainability reporting is a complex issue that should try to contemplate both the investor and stakeholder perspectives and highlights the existing trade-off and contrast between value to society and value to investors (La Torre et al., 2020, p. 715). This is opposed to the understanding of a so-called single materiality which would require information to be material from both perspectives—and not from either one or the other—to fall under the reporting obligation defined in the NFRD (Baumüller & Sopp, 2022).

The introduction of double materiality is seen as an "apparent deviation from the European Commission's previous approach of expanding the interpretation of "financial materiality" and confining the intended audience to investors" (Abhayawansa, 2022, p. 1372) to distinguish between the perspective of investors (who are served with "financial materiality") and of other stakeholders (whose needs are served by the adoption of "impact materiality").

The concept of double materiality for sustainability reporting is more explicitly adopted and regulated by the Corporate Sustainability Reporting Directive and the European Sustainability Reporting Standards (ESRSs). This outlines a definitive regulatory shift in the understanding of the nature and rationale of sustainability reporting in the European Union (Baumüller & Sopp, 2022). Indeed, the CSRD stresses that sustainability reporting has different users with different information needs and expectations:

> The first group of users consists of investors, including asset managers, who want to better understand the risks and opportunities that sustainability issues pose for their investments and the impacts of those investments on people and the environment. The second group of users consists of civil society actors, including non-governmental organisations and social partners, which wish to better hold undertakings to account for their impacts on people and the environment. Other stakeholders might also make use of sustainability information disclosed in annual reports, in particular, to foster comparability across and within market sectors. (par. 9)

By broadening the intended audience of sustainability reporting, the CSRD reinforces the notion of double materiality as a filter to determine what to include in a sustainability report, to serve both the interests of investors and other stakeholders,

Article 19a (1) of Directive 2013/34/EU as amended by the CSRD, states that companies concerned "shall include in the management report information necessary to understand the undertaking's impacts on sustainability matters, and information necessary to understand how sustainability matters affect the undertaking's development, performance and position".

This revised definition clarifies that that undertakings should consider each materiality perspective (impact and financial) in its own right, and should disclose information that is material from both perspectives as well as information that is material from only one perspective. The inherent assumption is that not all information necessary for stakeholders to hold companies accountable for their impacts is considered relevant for investors to understand sustainability-related risks and opportunities.

The European Sustainability Reporting Standards (ESRS) provide further principles and guidance for the interpretation and implementation of the double materiality under the CSRD.

The *ESRS 1 General Requirements* (European Commission, 2023) states that "The undertaking shall report on sustainability matters based on the double materiality principle" (par. 21) and clarifies that "Double materiality has two dimensions, namely: impact materiality and financial materiality" (par. 37), which are defined as follows:

- a sustainability matter is material from an impact perspective when "it pertains to the undertaking's material actual or potential, positive or negative impacts on people or the environment over the short-, medium- or long-term" (ESRS 1, par. 43). Impact materiality in the ESRSs is defined as in the GRI Standards;
- a sustainability matter is material from a financial perspective if "it triggers or could reasonably be expected to trigger material financial effects on the undertaking. This is the case when a sustainability matter generates risks or opportunities that have a material influence or could reasonably be expected to have a material influence, on the undertaking's development, financial position, financial performance, cash flows, access to finance, or cost of capital over the short-, medium- or long-term" (ESRS 1, par. 48).

Put differently, the double materiality approach intended to address the so-called outside-in perspective (risks and opportunities for the entity, "financial materiality") as well as the so-called inside-out perspective (positive and negative impacts of the entity, "impact materiality". ESRS 1 explicitly acknowledges the interdependencies between these materiality perspectives:

Impact materiality and financial materiality assessments are interrelated and the interdependencies between these two dimensions shall be considered. In general, the starting point is the assessment of impacts, although there may also be material risks and opportunities that are not related to the undertaking's impacts. A sustainability impact may be financially material from inception or become financially material, when it could reasonably be expected to affect the undertaking's financial position, financial performance, cash flows, access to finance, or cost of capital over the short-, medium- or long-term. (par. 38)

Further guidance to conduct a sustainability materiality analysis using the double materiality perspective is provided in the ESRS 1 Appendix Double Materiality (AR6-AR18).

Researchers raised several concerns over the concept of double materiality and discredited the arguments for having two different perspectives to determine material information. A survey of users of sustainability reports and follow-up interviews conducted by Jørgensen et al. (2022) revealed that respondents expressed a belief that is challenging to discern between financially material sustainability issues and those that are only material from the perspective. Furthermore, by creating a dichotomy, the double materiality perspective ignores the "rebound" and "boomerang" effects between the outside-in and inside-out perspectives (Giner & Luque-Vílchez, 2022) and the interrelationships between the two. Abhayawansa (2022, p. 1373) observes that while financial materiality is defined with reference to the information needs of a reasonable investor, the definition of impact materiality is not anchored on a clearly defined intended audience. The lack of a reference group or groups as the intended audience makes it difficult to identify what financially non-material sustainability topics are also not material from an impact perspective. Additionally, he observes that determining material sustainability information using double materiality is unlikely to help organizations and users appreciate that "Enterprise value and sustainability are intimately and reciprocally related and mutually reinforcing. In the normative sense, there cannot be sustainability-related risks and opportunities that could not reasonably be expected to affect an organisation's enterprise value" (p. 1375). Baumüller and Sopp (2022) highlight the risks associated with the incompatibility of certain parts of the new reporting requirements with the overall accounting framework in the EU, especially with regard to the necessary intersections with financial reporting in the management commentary, as well as the difficulty of determining materiality levels for ecological and social information per se, i.e., not considering financial impacts and the higher costs for the reporting companies, against very uncertain benefits. La Torre et al. (2020) argue that double materiality "may result in a risk management-oriented approach to stakeholder engagement that has nothing to do with broad corporate accountability to stakeholders" (p. 718) and will continue to privilege financial sustainability over social and environmental sustainability. They explain that the double materiality perspective poses the risk that "social and environmental materiality may be used only to assess social and environmental risks to preserve the company's financial value as companies continue to privilege financial sustainability over social and environmental sustainability" (La Torre et al., 2020, p. 715). De Villiers et al. (2022) further suggest that double materiality encloses the inscription of stakeholders' interests that are hard to balance and its trade-off may advantage the financial interests. Then, they argue that double materiality "may cause the financial capture of sustainability reporting, so reducing its broad accountability potential" (De Villiers et al., 2022, p. 738).

Dynamic Materiality

In any case, the different approaches to materiality should not be necessarily seen as opposite, but rather they form part of a coherent eco-system. In 2021, the GRI and SASB agreed that there is an opportunity to apply the two sets of standards in a complementary fashion to generate information that meets the needs of all constituencies and markets (GRI and SASB, 2021). The choice to combine and integrate GRI and SASB standards can be considered an attempt made by virtuous companies to enhance their accountability processes through the identification of alternative indicators (Pizzi et al., 2022).

To highlight how the two perspectives can interact, Giner and Luque-Vílchez (2022) refer to the "rebound" or "boomerang" effect, which suggests that "first the entity impacts people and the environment, and said impact can then rebound on its business model, subsequently affecting the entity's value".

As pointed out by Kuh et al. (2020, p. 13), "[a]s companies more rapidly change their business models, what is material to such companies will be changing in stride. Just as the new material topics will emerge for companies as the company evolves, some sustainability issues that were previously material financially to companies will no longer be". This characteristic of materiality, that the relative importance of sustainability issues to companies and their stakeholders as well as their sustainability impacts can vary over time, requires approaches to materiality that are not static (Jørgensen et al., 2022, p. 346).

To capture this shifting concept, the World Economic Forum introduced the new concept of "dynamic materiality". The main idea of dynamic materiality is that "what is financially immaterial to a company or industry today can become material tomorrow" (World Economic Forum, 2020, p. 5). The World Economic Forum (2020, p. 14) in the white paper entitled *Measuring Stakeholder Capitalism: Towards Common Metrics and Consistent Reporting of Sustainable Value Creation*, explains that "Materiality is a dynamic concept, in which issues once considered relevant only to social value can rapidly become financially material. Dynamic materiality indicates that the overlap between financial, social and environmental materiality can increase in the future. In this sense, sustainable value creation lies at the intersection of social and corporate value".

5.4   Corporate Approaches to Materiality: An Empirical Analysis

Research Design

While it is crucial for users to understand how the sustainability materiality principle is implemented, companies often provide disclosures that are both limited and poorly comparable in relation to the process of assessing material topics for sustainability reporting purposes (Abhayawansa & Adams, 2022; Beske et al., 2020; Unerman & Zappettini, 2014). Observed inconsistencies in materiality disclosures exist even among companies in the same sector, influenced by diverse factors and motivations that may not always align with the materiality definition used (Edgley et al., 2015; Farooq et al., 2021; Gerwanski et al., 2019; Jones et al., 2016; Steenkamp, 2018).

This section delves into the current state of corporate disclosure regarding materiality and the associated assessment process in sustainability reporting. It achieves this by scrutinizing and comparing the disclosure practices of the world's largest companies, offering additional insights to contribute to the existing debate on sustainability materiality disclosure.

In our research, we thoroughly examine how transparent companies are about applying the materiality principle in sustainability reporting. We specifically investigate whether companies provide comprehensive details about their materiality assessment process. In particular, our work is guided by three research questions:

RQ1: *To what extent is the materiality principle disclosed in the sustainability report?*

RQ2: *Which approach to materiality do companies disclose?*

RQ3: *To what extent do companies report on their materiality assessment process?*

Our empirical data for this research comprises materiality disclosures obtained from sustainability reports of a selected sample of European and US firms. The aim is to ensure a representative sample that includes companies adopting various sustainability reporting standards and employing diverse approaches to materiality. In the European context, our sample encompasses all companies listed in the EURO STOXX 50, the premier blue-chip index for the Eurozone, tracking the largest and most traded companies. The selection is based on the composition as of December 31, 2023. For the US, our sample includes all companies in the S&P 500 Top 50, representing the 50 largest companies from the S&P 500 and reflecting mega-cap performance in the US, also based on the composition as of December 31, 2023.

The sustainability disclosure for the financial year 2022, presented in various formats, was downloaded from the sampled companies' websites.

Consistent with previous studies (Fasan & Mio, 2017; Puroila & Mäkelä, 2019; Steenkamp, 2018), our research uses the content analysis technique (Krippendorff, 2004) to examine the quality and the extant of disclosure about the materiality assessment process. The main strength of the content analysis is that: "It provides a means for quantifying the contents of a text, and it does so by using a method that is clear and, in principle, repeatable by other researchers" (Denscombe, 1998, p. 168). The content analysis is conducted manually since our coding process requires a high degree of data interpretation and, therefore, cannot be automatized using software.

Our research involves four steps, as detailed in the following.

Step 1: Data collection (the basis for RQ1, RQ2, and RQ3)

Firstly, the sustainability reports undergo scrutiny. Additionally, by searching for keywords related to materiality, the relevant pages and contents covering materiality disclosure are identified and copied to form a dataset. The collected information about material issues and the processes of materiality determination will undergo further analysis.

Step 2: The use of the materiality principle (for RQ1)
In this step, we analyze materiality-related disclosures from the previous stage and identify additional statements where companies declare how the content of the report is determined. Our focus is on assessing how companies communicate the use of the materiality principle in their sustainability reports. Each report is then categorized as follows:

1. *No reference*: The materiality principle is not mentioned.
2. *Implicit reference*: Although not explicitly mentioned, the report refers to other principles or criteria for content determination that resonate with the materiality principle.
3. *Explicit reference*: The report explicitly acknowledges the use of the materiality principle to determine its content.

Step 3: Approach to materiality (for RQ2)

In the third step, we analyze and categorize the approach to materiality (impact, financial, or double materiality) that companies communicate using. To achieve this, we search for and collect statements defining the meaning of material topics and explaining the criteria used to assess and prioritize them.

Step 4: The level of disclosure about the materiality assessment process (for RQ3)
In the fourth step, our focus is on evaluating the extent of materiality disclosure to gauge the attention given to the materiality assessment process in sustainability reports. To assess the quality and quantity of materiality disclosure, we draw on insights from prior research (Farooq et al., 2021; Fasan & Mio, 2017; Ruiz-Lozano et al., 2022) and create a disclosure score, structured as follows:

- **Score 0: No reference.** This score is assigned to reports where the principle of materiality is not mentioned.

- **Score 1: Simply mentions.** A score of one is given when reporters state that materiality was a guiding principle in defining the report's contents without providing additional clarification.
- **Score 2: Brief discussion.** This score is assigned to reports without a dedicated section on materiality but briefly discusses initiatives taken to identify material issues.
- **Score 3: Process mention.** A score of three is given when reporters feature a dedicated section on materiality, mentioning the materiality analysis process without explicit reference to the steps undertaken. This section may or may not include a list of material issues.
- **Score 4: Process description.** A score of four is assigned to reports incorporating a dedicated section on materiality, including a list of material issues and a detailed description of the materiality assessment process, along with a brief discussion of each step.
- **Score 5: Detailed process analysis.** This highest score is allocated to reports where significant attention is devoted to the materiality issue. The report includes a specific section on materiality with a list of material topics and a comprehensive description of each step of the materiality analysis process.

The scoring system effectively captures the varying levels of detail in how companies communicate their approach to materiality in their sustainability reports.

Analysis of Results

The Use of the Materiality Principle in the Sustainability Report

With our first research question, we aim to assess the extent to which companies disclose the materiality principle in their sustainability reports.

Our findings indicate that companies in the S&P 50 and EURO STOXX 50 have incorporated and disclosed the materiality principle in their sustainability reports using three distinct approaches: "*no reference*", "*implicit reference*", and "*explicit reference*". A comparative analysis reveals significant differences between S&P 50 and EURO STOXX 50 companies (Table 5.3).

In general, a small proportion of companies—10% in the S&P 50 and 6% in the EURO STOXX 50—adopt a "no reference" approach. This

Table 5.3 The use of the materiality principle

	S&P 50 (%)	EURO STOXX 50 (%)
No reference	10	6
Implicit reference	52	2
Explicit reference	38	92

suggests that a limited number of sampled companies do not explicitly acknowledge the materiality principle or other criteria guiding the determination of content in their sustainability reports.

For instance, in explaining its approach to ESG and ESG reporting, Netflix only states that:

> Our Environmental, Social, and Governance (ESG) framework is informed by relevant reporting frameworks — including Sustainability Accounting Standard Board (SASB), Global Reporting Initiative (GRI) and Task Force on Climate-Related Financial Disclosures (TCFD). The Netflix Board of Directors oversees the Company's ESG efforts and receives regular updates from management in these areas. For this report, we focus on the following topics: Environment, Social and Governance. (Netflix ESG Report, 2022, p. 5)

In some instances, companies provide clarification that the criteria used to select information in their sustainability report differ from those employed for financial reporting purposes. For instance, Comcast states that:

> The inclusion of forward-looking and other statements in this Impact Report is not an indication that they are necessarily material to investors or required to be disclosed in our filings with the SEC. (Comcast 2023 Impact Report, p. 2)

In such instances, although users are informed that sustainability information may not be selected and communicated based on financial materiality, they are left unaware of the specific criteria used to filter sustainability information.

In the S&P 50, companies predominantly (52%) adopt the "implicit reference" approach, suggesting that a significant number incorporate the materiality principle without explicit terminology. In contrast, the EURO

STOXX 50 companies show a considerably lower percentage at 2%, indicating a strong inclination toward more explicit communication about materiality. Rather than explicitly recognizing the use of the materiality principle, these companies explain the process of defining the report's content using various terms or expressions that convey resonance with materiality since they indicate that sustainability topics and information undergo a selection process based on their importance.

Most companies assert that their sustainability report focuses on "ESG priorities" (or similar terms such as "sustainability priorities" and "priority ESG issues")—determined through a "prioritization assessment process"—while others use expressions like "most relevant", "most important", "most pertinent", or simply "key" topics.

For example, Eli Lilly and Company includes the following statement to explain the content and the boundaries of its online sustainability report:

> We believe transparency is important to ensuring accountability for our sustainability strategy, programs, and performance. We disclose relevant information and progress around the management of our ESG priorities and aim to stay up to date with relevant sustainability and social impact reporting regulations, frameworks, and standards that best meet the needs of our stakeholders.

Interestingly, more than half of US companies that adopt an "*implicit reference*" approach (14 out of 26) make explicit use of the terms "materiality" or "material" to specify that information selected for sustainability reporting should not be considered material as intended for financial reporting purposes. For instance, Apple clarifies that:

> The report does not cover all information about our business. References in this report to information should not be construed as a characterization regarding the materiality of such information to our financial results or for purposes of the US securities laws (Apple's 2022 ESG Report, p. 84).

In cases where the concept of materiality is explicitly employed solely to contrast the content of sustainability reporting with financial reporting, it suggests that these companies are aware of the materiality principle's significant application in financial reporting but they recognize the necessity to use other expressions, albeit conceptually similar to materiality, for sustainability reporting.

Companies adopting an "explicit reference" approach explicitly acknowledge that their sustainability report is centered on "ESG material topics" determined through a "materiality assessment process". This approach is prevalent among EURO STOXX 50 companies (92%), while S&P 50 companies exhibit a significantly lower percentage (38%). This suggests that European companies are more inclined to explicitly acknowledge the use of materiality in sustainability reporting compared to US companies.

For instance, AXA XL acknowledges that:

> Materiality is a comprehensive process that enables AXA XL, with our stakeholders, to understand the key ESG issues for our business. It helps us prioritize where our sustainability strategy should be focused and the actions we need to take. (AXA XL Sustainability Report 2022, p. 5)

Similarly, PepsiCo declares that:

> We report annually on the topics that were identified through our most recent assessment confirming the material sustainability issues that are addressed in our pep+ (PepsiCo Positive) strategy and that we believe represent PepsiCo's ESG performance. (PepsiCo 2022 GRI Index, p. 1)

In conclusion, our findings indicate that the application of the materiality principle is prevalent in the sustainability reporting of European companies, explicitly acknowledging a focus on material topics. Concerning the S&P 50 companies, our results suggest a tendency to incorporate a materiality filter in their sustainability reporting, often implicitly without explicitly using materiality-related terminology in the report. In sum, the use of sustainability materiality appears to be a well-established practice among European companies, while US companies, especially in terms of transparent communication about this principle, appear to be slightly behind. The differences between S&P 50 and EURO STOXX 50 companies in their approaches to materiality disclosure could stem from the different regulatory sustainability reporting requirements or varying stakeholder expectations.

The Disclosed Approach to Materiality

Our second research aims to investigate the approach to materiality that companies disclose.

This analysis is limited to the 45 companies within the S&P 50 and the 47 companies within EURO STOXX 50 that, explicitly or implicitly, acknowledge adopting the materiality principle in their sustainability reporting. In the absence of explicit and clear references outlining the use of a specific materiality approach (impact, financial, or double materiality), we based our analysis on statements explaining the criteria used for materiality assessment and prioritization of material topics.

Table 5.4 provides some quotations from the sampled companies' sustainability reports serving as examples for each of the identified approaches.

As shown by Table 5.5, our findings suggest that financial materiality is the predominant approach in the S&P 50 sample (47%), while double materiality is prevalent among companies within the EURO STOXX 50 (36%).

Results indicate that 47% of S&P 50 companies and 9% of EURO STOXX 50 companies conduct materiality assessments from both the "inside-out" (i.e., impacts of corporate activities on sustainability issues or stakeholders) and "outside-in" (i.e., impacts of sustainability issues on the business success) perspectives, but they require information to be material from both perspectives—and not from either the one or the other—to be prioritized for sustainability reporting purposes. In the literature, this approach, known as "single materiality", is traced back to financial materiality (Baumüller & Sopp, p. 18), since it requires sustainability matters to have financial implications for the company to be considered as material.

The impact materiality perspective is adopted by a similar number of companies in the two samples (18% in the S&P 50 and 21% in the EURO STOXX 50). The double materiality approach is adopted by a minority of companies in the S&P 50 (7%) and the majority of companies in the EURO STOXX 50 (36%), which anticipate the requirements of the CSRD and the ESRSs. These companies consider material those topics

Table 5.4 The disclosed approach to sustainability materiality

	S&P 50 (%)	EURO STOXX 50 (%)
Single/financial materiality	47	9
Vague or ambiguous	29	34
Impact materiality	18	21
Double materiality	7	36

Table 5.5 Examples of disclosed approaches to materiality

Approach	Quotations S&P 50	Quotations EURO STOXX 50
Single/financial materiality	"Most recently conducted in 2020, our periodic materiality assessment process helps us determine where to focus ESG efforts. From this process, we have identified six ESG-oriented "material drivers" that reflect high importance to our stakeholders and generate high impact for AbbVie" (AbbVie 2022 ESG Action Report, p. 12) "This assessment is designed to allow us to identify and prioritize the ESG issues that are of greatest concern to our stakeholders and that impact the success of our business" (Intel 2022–2023 Corporate Responsibility Report, p. 27) "Our material issues articulate what matters most to our business and our stakeholders. This awareness is crucial to identify and manage our risks and opportunities, and to respond effectively to our stakeholders" (2020 Microsoft Sustainability Report, p. 82)	"The core topics identified were then assessed in terms of their Double materiality for BASF. Each sustainability aspect was Considered from two perspectives: To assess sustainability relevance ("impact materiality"), both the actual and the potential positive and negative impacts of our company's activities were considered along three stages of the value chain (upstream, own operations, downstream). Here, we assessed the scale of impact, its scope, and likelihood of occurrence. The individual topics were classified based on their potential financial impacts on BASF as part of the financial materiality analysis [...] A sustainability aspect is considered material in the sense of double materiality if it has been classified as material both at the level of sustainability relevance and at the level of financial relevance" (BASF Report 2022, p. 46)

(continued)

Table 5.5 (continued)

Approach	Quotations S&P 50	Quotations EURO STOXX 50
Vague or ambiguous	"Since 2013, we have regularly engaged third party experts to conduct ESG materiality assessments, which identify and prioritize the corporate responsibility impacts, risks and opportunities that we address to help ensure our long-term business success" (2022 Qualcomm Corporate Responsibility Report, p. 9) "Boeing considers stakeholders' interests to identify and prioritize the most relevant issues and to assess the most significant challenges and risks facing the company" (2023 Boeing Sustainability Report, p. 11) "We continually assess our ESG priorities based on their importance to our business and our stakeholders" (Accenture 360 Value Report 2022, p. 86)	"In reporting terms, materiality relates to the extent to which measures have a significant impact on an organization and its ability to create financial and non-financial value, for itself and its stakeholders" (Saint-Gobain Universal Registration Document 2022, p. 54) "The materiality analysis is used to identify and evaluate the most important sustainability issues for the Group. Based on the business model and its impact on society, the focus is on key ESG requirements, stakeholder expectations, and compliance with legal requirements and internationally established reporting standards" (Volkswagen Group Sustainability Report 2022, p. 18)
Impact materiality	"McDonald's prioritizes environmental and social issues where we can have the greatest impact and are most important to our stakeholders" (McDonald's Corporation Purpose & Impact Report 2022–2023, p. 8) "After establishing our sustainability goals in 2021, we saw a need and an opportunity to establish a framework around the areas where Wells Fargo can create the greatest impact" (Wells Fargo Sustainability & Governance Report 2023, p. 6)	"The assessment was performed as a four-step process in accordance with the GRI 2021 standards to assess actual and potential negative and positive impacts upon the economy, environment, and people, including impacts on their human rights" (Santander Consumer Bank AS Sustainability Report 2022, p. 10) "The materiality analysis has been implemented in accordance with the GRI Standards. Our Ferrari Leadership Team (FLT) was involved, through one-to-one interviews, in identifying and prioritizing our most relevant impacts on the economy, environment, and people, including impacts on human rights, across our activities and business relationships" (Ferrari N.V. Sustainability Report 2022, p. 38)

Approach	Quotations S&P 50	Quotations EURO STOXX 50
Double materiality	"Our last Priority Topics Assessment (PTA) was conducted in 2021 with a double materiality focus, by which we examined ESG priority topics from two standpoints: the impact of a topic on Johnson & Johnson's business results and the impact of Johnson & Johnson's business on people, the environment and society in general. Our Priority Topics Matrix presents our ESG priorities" (Johnson & Johnson, 2022, Health for Humanity Report, p. 8)	"In 2022, Sanofi conducted a double materiality assessment with support from an independent third party. This covered the impacts of our activities on society (impact materiality), and impacts that societal changes might have on Sanofi's performance (financial materiality)" (Sanofi, 2022, Universal Registration Document, p. 5) "In 2022, L'Oréal conducted a double materiality analysis to anticipate future European regulatory requirements currently in preparation and to continue its dialogue with stakeholders. This analysis was carried out by an external third party and is based on the double materiality principle established by the Non-Financial Reporting Directive and fully enshrined in the Corporate Sustainability Reporting Directive (CSRD). It was developed on the basis of the draft sustainability reporting standards submitted for consultation in April 2022 (ESRS). These standards define materiality in relation to two different dimensions: financial materiality (the impact of sustainability issues on the development, position or financial performance of a company); and impact materiality (the impact of a company on people or the environment) (L'Oréal Universal Registration Document, 2022, p. 155)

reflecting either significant economic, environmental, and social impacts (inside-out perspective) or significant impacts of sustainability topics on the business success (outside-in perspective).

Finally, a substantial number of companies (29% in the S&P 50 and 34% in the EURO STOXX 50) disclose unclear or vague statements about the criteria used to conduct the materiality assessment analysis, making it impossible for users to understand the materiality approach used. In some of these cases, the reporting company mentions the use of a double perspective, but how the financial and impact logics are combined (i.e., whether material topics are those topics relevant from both perspectives or only one of them) is not explained.

Extant of Disclosure on the Materiality Assessment Process
To address our third research question, we assess the quality of disclosure on the materiality assessment process. This analysis offers valuable insights into the current state of materiality disclosure among the S&P 50 and the EURO STOXX 50 companies, providing a basis for further analysis, benchmarking, and potential improvements in sustainability reporting practices. Overall, our results (see Table 5.6) indicate substantial diversity in reporting practices among the US and European largest companies, with varying levels of detail and transparency about materiality assessments.

Interestingly, the average materiality disclosure score is relatively low and similar in the two samples: 2.36 for the S&P 50 and 2.80 for the EURO STOXX 50. This suggests a general hesitancy among companies to offer comprehensive disclosure regarding their materiality assessment processes. Notably, European companies appear to be slightly more

Table 5.6 Level of disclosure on the materiality assessment process

Level of disclosure	S&P 50 (%)	EURO STOXX 50 (%)
No reference	10	6
Simply mentions	18	10
Brief discussion	18	12
Process mention	36	44
Process description	16	24
Detailed process analysis	2	4

transparent than their US counterparts in communicating materiality analyses.

Specifically, apart from companies that do not even mention the principle of materiality—10% in S&P 50 and 6% in the EURO STOXX 50—three approaches to disclosure can be identified:

- *Limited Disclosure* (Simply mentions, Brief discussion). A combined 36% of the S&P 50 sample and 22% of the EURO STOXX 50 sample fall into categories where disclosure on the materiality assessment process is absent or relatively limited. This suggests that nearly half of the companies may not be providing sufficient information to allow users to understand how they determine material issues for their sustainability reports.
- *Moderate Disclosure* (Process mention). 36% of the S&P 50 sample and 44% of the EURO STOXX 50 sample include companies that mention the materiality analysis process and briefly describe the main initiatives carried out for materiality assessments but do not provide explicit details on the steps of the materiality process. This suggests a common practice of acknowledging materiality without delving into a comprehensive explanation of its assessment.
- *Detailed Disclosure* (Process description, Detailed process analysis). A total of 18% of companies in the S&P 50 sample and 28% of companies in the EURO STOXX 50 sample fall into categories involving more detailed disclosure. These companies provide a specific materiality section with a list of material issues and offer consistent degrees of information about the materiality assessment process and a brief description of each step (typically: identification, assessment, prioritization, and validation).

There is only one US company and two European companies that provide a detailed process analysis and that stand out as exemplars in transparency, potentially serving as benchmarks for best practices in materiality disclosure.

5.5   Concluding Remarks

Materiality serves as the cornerstone of sustainability reporting, shaping the entirety of the reporting process. Our emphasis on materiality has led to two main findings.

First, our analysis reveals that materiality remains a subject of debate within sustainability reporting standards and frameworks, with three distinct approaches prevalent: impact, financial, and double materiality. This diversity of perspectives underscores the lack of convergence in the sustainability reporting landscape. While this divergence is not inherently negative, it underscores the need for companies to transparently disclose their chosen approach to materiality in their reports. This transparency enables stakeholders to understand the criteria utilized in identifying material topics and facilitates meaningful comparisons across companies. Second, our empirical analysis suggests that there appears to be room for improvement in the quality of materiality disclosure across the spectrum, from companies providing minimal information to those with brief discussions and even among those acknowledging the materiality process without providing explicit details on the steps.

This analysis encourages companies to reflect on their reporting practices, considering the expectations of stakeholders and the evolving landscape of sustainability reporting. As stakeholders increasingly prioritize sustainability information, there might be a growing expectation for companies to provide detailed and comprehensive disclosure on materiality assessments. Additionally, regulatory bodies may consider revisiting guidelines to encourage more standardized and thorough reporting practices.

As with any research, our empirical study is not without its limitations. Firstly, the scope of our sample is confined to companies listed on EURO STOXX 50 and S&P 50, representing some of the largest entities globally. Consequently, the generalizability of our findings to a broader spectrum of companies may be limited. Secondly, due to the nature of our research design, which is cross-sectional, our study is constrained by the collection of data from a single year. This limitation prevents us from capturing potential variations or trends over time, and a longitudinal approach could provide a more comprehensive understanding of the dynamics involved. Thirdly, it is essential to acknowledge that our study does not delve into the examination of the determinants or consequences associated with varying levels of transparency in materiality assessments.

References

Abhayawansa, S. (2022). Swimming against the tide: Back to single materiality for sustainability reporting. *Sustainability Accounting, Management and Policy Journal, 13*(6), 1361–1385.

Abhayawansa, S., & Adams, C. (2022). Towards a conceptual framework for non-financial reporting inclusive of pandemic and climate risk reporting. *Meditari Accountancy Research, 30*(3), 710–738.

Baumüller, J., & Schaffhauser-Linzatti, M.-M. (2018). In search of materiality for nonfinancial information—Reporting requirements of the Directive 2014/95/EU. *NachhaltigkeitsManagementForum | Sustainability Management Forum, 26*(1–4), 101–111.

Baumüller, J., & Sopp, K. (2022). Double materiality and the shift from non-financial to European sustainability reporting: Review, outlook and implications. *Journal of Applied Accounting Research, 23*(1), 8–28.

Bernstein, L. A. (1967). The concept of materiality. *The Accounting Review, 42*(1), 86–95.

Beske, F., Haustein, E., & Lorson, P. C. (2020). Materiality analysis in sustainability and integrated reports. *Sustainability Accounting, Management and Policy Journal, 11*(1), 162–186.

Bolt, R., & Tregidga, H. (2023). Methodological insights "materiality is …": Sensemaking and sensegiving through storytelling. *Accounting, Auditing & Accountability Journal, 36*(1), 403–427.

Brennan, N., & Gray, S. (2005). The impact of materiality: Accounting's best kept secret. *Asian Academy of Management Journal of Accounting and Finance, 1*, 1–31.

Calabrese, A., Costa, R., & Rosati, F. (2015). A feedback-based model for CSR assessment and materiality analysis. *Accounting Forum, 39*(4), 312–327.

Canning, M., O'Dwyer, B., & Georgakopoulos, G. (2019). Processes of auditability in sustainability assurance—The case of materiality construction. *Accounting and Business Research, 49*(1), 1–27.

Carvajal, M., & Nadeem, M. (2023). Financially material sustainability reporting and firm performance in New Zealand. *Meditari Accountancy Research, 31*(4), 938–969.

Cerbone, D., & Maroun, W. (2020). Materiality in an integrated reporting setting: Insights using an institutional logics framework. *The British Accounting Review, 52*(3), 100876.

Chong, H. G. (2015). A review on the evolution of the definitions of materiality. *International Journal of Economics and Accounting, 6*(1), 15.

Christensen, H. B., Hail, L., & Leuz, C. (2021). Mandatory CSR and sustainability reporting: Economic analysis and literature review. *Review of Accounting Studies, 26*(3), 1176–1248.

Clark, C. E. (2021). How do standard setters define materiality and why does it matter? *Business Ethics, the Environment & Responsibility, 30*(3), 378–391.

Consolandi, C., Eccles, R. G., & Gabbi, G. (2022). How material is a material issue? Stock returns and the financial relevance and financial intensity of ESG materiality. *Journal of Sustainable Finance & Investment, 12*(4), 1045–1068.

Cooper, S., & Michelon, G. (2022). Conceptions of materiality in sustainability reporting frameworks: Commonalities, differences and possibilities. In C. Adams (A c. Di), *Handbook of accounting and Sustainability* (pp. 44–66). Edward Elgar Publishing.

Corporate Reporting Dialogue. (2016). *Statement of common principles of materiality of the corporate reporting dialogue.* http://sirse.info/wp-content/uploads/2016/04/Statement-of-Common-Principles-of-Materiality.pdf

Deegan, C., & Rankin, M. (1997). The materiality of environmental information to users of annual reports. *Accounting, Auditing & Accountability Journal, 10*(4), 562–583.

Denscombe, M. (1998). *The good research guide for small-scale social research project.* Open University Press.

De Villiers, C., La Torre, M., & Molinari, M. (2022). The Global Reporting Initiative's (GRI) past, present and future: Critical reflections and a research agenda on sustainability reporting (standard-setting). *Pacific Accounting Review, 34*(5), 728–747.

Eccles, R. G., Krzus, M. P., Rogers, J., & Serafeim, G. (2012). The need for sector-specific materiality and sustainability reporting standards. *Journal of Applied Corporate Finance, 24*(2), 65–71.

Edgley, C. (2014). A genealogy of accounting materiality. *Critical Perspectives on Accounting, 25*(3), 255–271.

Edgley, C., Jones, M. J., & Atkins, J. (2015). The adoption of the materiality concept in social and environmental reporting assurance: A field study approach. *The British Accounting Review, 47*(1), 1–18.

European Commission. (2017). *Communication from the commission. Guidelines on non-financial reporting (methodology for reporting non-financial information) (2017/C 215/01).* https://eur-lex.europa.eu/legal-content/EN/TXT/PDF/?uri=CELEX:52017XC0705(01)

European Commission. (2019). *Communication from the commission. Guidelines on non-financial reporting: Supplement on reporting climate-related information (2019/C 209/01).* https://eur-lex.europa.eu/legal-content/EN/TXT/PDF/?uri=CELEX:52019XC0620(01)

Farooq, M. B., Zaman, R., Sarraj, D., & Khalid, F. (2021). Examining the extent of and drivers for materiality assessment disclosures in sustainability reports. *Sustainability Accounting, Management and Policy Journal, 12*(5), 965–1002.

Fasan, M., & Mio, C. (2017). Fostering stakeholder engagement: The role of materiality disclosure in integrated reporting. *Business Strategy and the Environment, 26*(3), 288–305.

Financial Accounting Standards Board (FASB). (1975). *Discussion memorandum: Criteria for determining materiality.* FASB.

Financial Accounting Standards Board (FASB). (2018). Statement of Financial Accounting Concepts No. 8. Conceptual Framework for Financial Reporting. *Chapter 3, Qualitative Characteristics of Useful Financial Information.*

Fiandrino, S., Tonelli, A., & Devalle, A. (2022). Sustainability materiality research: A systematic literature review of methods, theories and academic themes. *Qualitative Research in Accounting & Management, 19*(5), 665–695.

Gerwanski, J., Kordsachia, O., & Velte, P. (2019). Determinants of materiality disclosure quality in integrated reporting: Empirical evidence from an international setting. *Business Strategy and the Environment, 28*(5), 750–770.

Giner, B., & Luque-Vílchez, M. (2022). A commentary on the "new" institutional actors in sustainability reporting standard-setting: A European perspective. *Sustainability Accounting, Management and Policy Journal, 13*(6), 1284–1309.

Global Reporting Initiative (GRI). (2021a). *GRI 1: Foundation 2021.* https://www.globalreporting.org/how-to-use-the-gri-standards/gri-standards-english-language/

Global Reporting Initiative (GRI). (2021b). *GRI 3 Material Topics 2021.* https://www.globalreporting.org/how-to-use-the-gri-standards/gri-standards-english-language/

Global Reporting Initiative (GRI) and Sustainability Accounting Standards Board (SASB). (2021). *A practical guide to sustainability reporting Using GRI and SASB Standards.* https://www.globalreporting.org/media/mlkjpn1i/gri-sasb-joint-publication-april-2021.pdf

Global Reporting Initiative (GRI). (2022). GRI Universal Standards 2021. Frequently Asked Questions (FAQs).

Holstrum, G. L., & Messier, W. F., Jr. (1982). A review and integration of empirical research on materiality. *Auditing: A Journal of Practice & Theory, 2*(1), 45–63.

Hsu, C.-W., Lee, W.-H., & Chao, W.-C. (2013). Materiality analysis model in sustainability reporting: A case study at Lite-On Technology Corporation. *Journal of Cleaner Production, 57*, 142–151.

International Accounting Standards Board (IASB). (2018). *Conceptual framework for financial reporting.* https://www.ifrs.org/content/dam/ifrs/publications/pdf-standards/english/2021/issued/part-a/conceptual-framework-for-financial-reporting.pdf

International Sustainability Standards Board (ISSB). (2023). *IFRS S1 General requirements for disclosure of sustainability-related financial information*. https://www.ifrs.org/issued-standards/ifrs-sustainability-standards-navigator/ifrs-s1-general-requirements/#about

International Integrated Reporting Council (IIRC). (2021). *International <IR> framework*. https://integratedreporting.ifrs.org/wp-content/uploads/2021/01/InternationalIntegratedReportingFramework.pdf

Jones, P., Comfort, D., & Hillier, D. (2016). Materiality in corporate sustainability reporting within UK retailing: Materiality in sustainability reporting and UK retailers. *Journal of Public Affairs, 16*(1), 81–90.

Jørgensen, S., Mjøs, A., & Pedersen, L. J. T. (2022). Sustainability reporting and approaches to materiality: Tensions and potential resolutions. *Sustainability Accounting, Management and Policy Journal, 13*(2), 341–361.

Khan, M., Serafeim, G., & Yoon, A. (2016). Corporate sustainability: First evidence on materiality. *The Accounting Review, 91*(6), 1697–1724.

Krippendorff, K. (2004). *Content Analysis: An introduction to its methodology*. Sage.

Kuh, T. Shepley, A. Bala, G., & Flowers, M. (2020). *Dynamic materiality: Measuring what matters*. https://ssrn.com/abstract=3521035

La Torre, M., Sabelfeld, S., Blomkvist, M., & Dumay, J. (2020). Rebuilding trust: Sustainability and non-financial reporting and the European Union regulation. *Meditari Accountancy Research, 28*(5), 701–725.

Lai, A., Melloni, G., & Stacchezzini, R. (2017). What does materiality mean to integrated reporting preparers? An empirical exploration. *Meditari Accountancy Research, 25*(4), 533–552.

Machado, B. A. A., Dias, L. C. P., & Fonseca, A. (2021). Transparency of materiality analysis in GRI—Based sustainability reports. *Corporate Social Responsibility and Environmental Management, 28*(2), 570–580.

Messier, W. F., Martinov-Bennie, N., & Eilifsen, A. (2005). A review and integration of empirical research on materiality: Two decades later. *Auditing: A Journal of Practice & Theory, 24*(2), 153–187.

Mio, C., Fasan, M., & Costantini, A. (2019). Materiality in integrated and sustainability reporting: A paradigm shift? *Business Strategy and the Environment, 29*(1), 306–320.

Moroney, R., & Trotman, K. T. (2016). Differences in auditors' materiality assessments when auditing financial statements and sustainability reports. *Contemporary Accounting Research, 33*(2), 551–575.

Perera-Aldama, L. (2023). GRI and materiality: Discussions and challenges. *Sustainability Accounting, Management and Policy Journal, 14*(4), 884–903.

Pizzi, S., Principale, S., & de Nuccio, E. (2022). Material sustainability information and reporting standards. Exploring the differences between GRI and SASB. *Meditari Accountancy Research, 31*(6), 1654–1674.

Puroila, J., & Mäkelä, H. (2019). Matter of opinion: Exploring the socio-political nature of materiality disclosures in sustainability reporting. *Accounting, Auditing & Accountability Journal, 32*(4), 1043–1072.

Reimsbach, D., Schiemann, F., Hahn, R., & Schmiedchen, E. (2020). In the eyes of the beholder: Experimental evidence on the contested nature of materiality in sustainability reporting. *Organization & Environment, 33*(4), 624–651.

Roberts, R. W., & Dwyer, P. D. (1998). An Analysis of materiality and reasonable assurance: Professional mystification and paternalism in auditing. *Journal of Business Ethics, 17*, 569–578.

Ruiz-Lozano, M., De Vicente-Lama, M., Tirado-Valencia, P., & Cordobés-Madueño, M. (2022). The disclosure of the materiality process in sustainability reporting by Spanish state-owned enterprises. *Accounting, Auditing & Accountability Journal, 35*(2), 385–412.

Schiehll, E., & Kolahgar, S. (2021). Financial materiality in the informativeness of sustainability reporting. *Business Strategy and the Environment, 30*(2), 840–855.

Steenkamp, N. (2018). Top ten South African companies' disclosure of materiality determination process and material issues in integrated reports. *Journal of Intellectual Capital, 19*(2), 230–247.

Stolowy, H., & Paugam, L. (2023). Sustainability reporting: Is convergence possible? *Accounting in Europe, 20*(2), 139–165.

Sustainability Accounting Standards Board (SASB). (2017). *SASB conceptual framework.* https://www.sasb.org/wp-content/uploads/2019/05/SASB-Conceptual-Framework.pdf?source=post_page

Task Force on Climate-related Financial Disclosures (TCFD). (2017). *Recommendations of the task force on climate-related financial disclosures.* https://www.fsb-tcfd.org/recommendations/

Torelli, R., Balluchi, F., & Furlotti, K. (2019). The materiality assessment and stakeholder engagement: A content analysis of sustainability reports. *Corporate Social Responsibility and Environmental Management, 27*(2), 470–484.

Unerman, J., & Zappettini, F. (2014). Incorporating materiality considerations into analyses of absence from sustainability reporting. *Social and Environmental Accountability Journal, 34*(3), 172–186.

Whitehead, J. (2017). Prioritizing sustainability indicators: Using materiality analysis to guide sustainability assessment and strategy. *Business Strategy and the Environment, 26*(3), 399–412.

World Economic Forum. (2020). *Measuring stakeholder capitalism. towards common metrics and consistent reporting of sustainable value creation.* https://www.weforum.org/publications/measuring-stakeholder-capitalism-towards-common-metrics-and-consistent-reporting-of-sustainable-value-creation/

CHAPTER 6

Conclusion

Abstract The concluding chapter serves as a comprehensive summary of the book, offering a discussion of significant insights and contributions to both research and practice. It systematically highlights the primary benefits derived from the book, emphasizing its pivotal role in enhancing comprehension of the ongoing academic discourse on sustainability reporting and tracking the evolutionary trajectory of regulation. Additionally, the section delves into the research implications, presenting potential avenues for further exploration and scholarly inquiry. It extends its focus to the practical applications of the insights, underlining their relevance in real-world corporate settings. The chapter concludes by providing insightful suggestions for the future evolution of sustainability reporting practices and research, serving as a roadmap for continued advancements in this dynamic field.

Keywords Sustainability reporting · Non-financial disclosure · Sustainability accounting · Corporate reporting

6.1   FUTURE DEVELOPMENT AND DIRECTIONS FOR SUSTAINABILITY REPORTING

The landscape of sustainability reporting and its regulation has undergone significant changes and evolution in recent years, with the European Union emerging as a frontrunner in this field. The European Commission has taken proactive steps by issuing various regulatory measures, such as recommendations and communications, aimed at encouraging companies to voluntarily disclose non-financial information and adopt models of socio-environmental reporting. However, in response to unsatisfactory outcomes, the European Commission transitioned toward implementing hard-law regulations, notably the Non-Financial Reporting Directive (NFRD) in 2014 and the Corporate Sustainability Reporting Directive (CSRD) in 2022.

The CSRD represents the most advanced form of mandatory sustainability reporting regulation. Building upon the requirements of the NFRD, the CSRD extends the scope of mandatory reporting to encompass all large companies and those SMEs listed on regulated markets (excluding micro-enterprises). It mandates the audit (assurance) of reported information, introduces more comprehensive reporting standards, including the application of mandatory EU sustainability reporting standards issued by the European Financial Reporting Advisory Group (EFRAG), and requires the digital tagging of reported information. The European Commission emphasizes the importance of providing investors with adequate information on the sustainability performance of listed companies and aims to prevent large companies from being excluded from investment portfolios due to a lack of sustainability disclosure. In comparison with other regions such as the UK, the USA, Australia, New Zealand, India, Asia, and South Africa, the CSRD places the European Union significantly ahead in the regulation of sustainability reporting.

As observed in the realm of financial reporting, sustainability reporting landscape has been also characterized by an increased tension (yet to be resolved) between the need for international accounting harmonization and the development of national initiatives. Having a plethora of standards-setting bodies and the lack of uniform mandatory requirements across different jurisdictions pose significant challenges to achieving global convergence in sustainability reporting (Stolowy & Paugam, 2023) and threatens informational comparability. This is evident from the 2024 edition of the IFAC study "The State of Play: Sustainability Disclosure and Assurance", which provides a thorough global assessment of the application of sustainability reporting standards and frameworks. According to this investigation, the GRI Standards continue to be the most widely used standard. When examining specific countries, it becomes evident that almost all European reporting entities (e.g., 100% in Italy and 98% in Spain) adopt the GRI Standards, while their application seems much less widespread in the UK, South Africa and the USA. The most commonly implemented reporting standards in the USA are the TCFD Recommendations and the SASB. The state of confusion is aggravated by the fact that in some countries sustainability reporting regulation is limited to specific areas of disclosure (such as environment and climate change concerns in the US), preventing comprehensive sustainability reporting at a time when stakeholders increasingly expect information also on social issues, such as diversity and equity, and governance. Having a lack of convergence can have adverse consequences, such as decrease the legitimacy of each reporting standard, increase disclosure processing costs, complexify the integration of sustainability information into investment decisions, and create regulatory arbitrage opportunities for companies (Stolowy & Paugam, 2023). Then, further work needs to be conducted by regulators and standards-setting bodies to make sustainability reporting converge, by reducing the state of confusion around the conceptualization of sustainability and the diversity in the objectives and the contents of reporting requirements.

Further critical issues in sustainability reporting emerge from our systematic literature review, which reveals a spectrum of themes that are essential to understand the impact of the regulatory evolution on non-financial disclosure and the specific theories that underpin this debate. First, the designation given to different forms of reporting and the location deemed preferable for the disclosure of NFI represents relevant discussed matters. Applying a standard designation to disclosures concerning sustainability issues proved to be a significant challenge. The heading "social and environmental reporting" was used at first, but in more recent times, it has become less common (Farneti & Guthrie, 2009), progressively making way for designations like "corporate social responsibility reports", "non-financial reporting", and "sustainability reporting" (Adams & Larrinaga-González, 2007; Stolowy & Paugam, 2018). Furthermore, the different headings were also meant to indicate the content of the reports (intended mainly to convey socio-environmental issues), to emphasize the type of corporate communication (in the absence of specific regulatory guidance), and to point out the differentiation from the actual corporate financial reporting (Thorne et al., 2014). Whenever the designation "sustainability reporting" initially emerged, it appeared to be considered very broadly, lacking an exact meaning or standard approach. Apart from the report's heading, there has also been debate in the literature on the communication tool to be employed for non-financial disclosure. On the one hand, according to a line of research, even for NFI, the annual financial report serves as a highly effective communication medium (Farneti & Guthrie, 2009). On the other hand, long-standing research has maintained that non-financial disclosure relies on annual stand-alone documents (Cho et al., 2015b; Michelon et al., 2015; Milne & Gray, 2007; Thorne et al., 2014). Second, the literature has long emphasized the need to establish a process aimed at determining the materiality of disclosed sustainability issues to increase the transparency of information, the quality of reports, and the trust of stakeholders (Beske et al., 2020; Dilling & Harris, 2018; Eccles et al., 2012; Fasan & Mio, 2017; Steenkamp, 2018). The identification of material information to be disclosed is based on the information available to the organization, stakeholder expectations, corporate and industry risks and opportunities, and the various reports disseminated by the company (Ruiz-Lozano et al., 2022), as recalled and explored in the third chapter. Third, there are a multitude of motivations that can drive a company to report (even voluntarily) on sustainability issues. The current concerns, particularly those related to the environmental and social spheres, constitute a real and visible threat to worldwide stability. Human beings must try to alter their path by considering the present and future effects of their actions since they are both participants in and witnesses of the actual global evolution. Given the urgent and increasing need for information in this regard, companies should appropriately answer to their stakeholders by describing and explaining their efforts through sustainability reporting, which is the primary source for achieving this (Milne et al., 2009). Furthermore, as business activity is legitimated by society, the latter commits to preserve the former. Companies exhibit an ethical responsibility toward society, from which they need to obtain legitimacy to keep going. The role and motivation of sustainability reporting have been extensively explored using several theoretical frameworks (i.e., the theory of voluntary disclosure, signaling theory, legitimacy theory, stakeholder theory, and institutional theory). Fourth, applying the recalled and analyzed theoretical frameworks, some studies considered and tested the relationship between non-financial disclosure and corporate performance. The challenge of determining a relationship between information disclosure and effective sustainability performance has long been addressed and discussed in the literature, recalling also the concepts of "organized hypocrisy" and "organizational facades" (Cho et al., 2015a). Fifth, rather than focusing just on quantitative expectations, the literature emphasizes the relevance of focusing on the completeness and quality of NFI. The examination of reports' informational quality is especially crucial since, as a prominent stream of research has suggested, there may be a significant relationship between this quality and the value attributed to the company (Barth et al., 2017). Additionally, the majority of pertinent research has been done in the private sector: the public sector has contributed far fewer papers to the issue under discussion. Sixth, issues about the reliability and the quality of non-financial disclosure have prompted a rise in auditing procedures in recent years to determine whether companies' actual performance is based on their reported information (García-Sánchez et al., 2022; Kim et al., 2019). This also has an important connection to the concept defined as the "expectation gap", which is the discrepancy between the real and expected benefits of assurance. It relates to the distinction between assurance as perceived and as provided, highlighting the importance of the discussions surrounding the categories of auditing activities (limited or reasonable assurance, e.g.) and the degree of accuracy and completeness with which such activities ought to be conducted (Farooq and De Villiers, 2020; Maroun, 2019).

Additional issues are related to the cornerstone of the sustainability reporting process, which is sustainability materiality: connections and differences with materiality in financial reporting; the importance of materiality for stakeholder engagement; the characteristics of the materiality assessment process, and the determinants of its quality; the disclosure of the materiality assessment process and its determinant; the relationship between the quality of materiality assessment and its disclosure and the financial performance of the firms. Our empirical research on the quality and extent of corporate disclosure on sustainability materiality assessments provides relevant contributions to current academic knowledge and practice on the issue of materiality. First, our study adds to prior literature investigating the materiality disclosure practices of companies (Beske et al., 2020; Farooq et al., 2021; Fasan & Mio, 2017; Puroila & Mäkelä, 2019; Steenkamp, 2018). We corroborate existing studies revealing the reluctance of the largest companies in the world to disclose materiality assessments (Machado et al., 2021; Ruiz-Lozano et al., 2022). Our results reflect a spectrum of disclosure practices, from minimal mentions of materiality to more comprehensive reporting that delves into the specific steps of the materiality assessment process. More specifically, our study yields three main insights. Firstly, we explore the contrast between European and American companies in using and acknowledging the materiality principle in sustainability reporting. While the majority of EURO STOXX 50 companies explicitly recognize and embrace the importance of the materiality principle in their sustainability reports, S&P 50 counterparts tend to incorporate conceptually similar reporting principles and criteria without explicitly mentioning materiality, except to exclude the relevance of sustainability information for financial reporting purposes. Moving on to our second point, our research sheds light on the diversity in the materiality approaches adopted by companies. European companies, especially in anticipation of the forthcoming CSRD and ESRS, lean toward embracing the double materiality logic. On the other hand, American companies seem to gravitate more toward applying financial materiality. Notably, approximately one-third of the companies utilize vague and ambiguous expressions, creating challenges for users in understanding the specifics of their materiality approach. In conclusion, our investigation underscores limited transparency across both European and American companies when it comes to disclosing materiality assessment processes. Notably, the steps involved in the materiality analysis are not adequately presented to users, raising concerns about the reliability and trustworthiness of the reports. This is crucial since the outcomes of the materiality analysis fundamentally shape the content of the entire sustainability report. Addressing this transparency gap is essential to maintain stakeholder trust and confidence in the credibility of these reports. In terms of practical implications, offering insights into the quality and extent of materiality disclosure can pave the way for the development of comprehensive guidelines for best reporting practices. The findings of this research hold the potential to offer valuable insights to standard setters, ensuring the establishment of higher standards of quality, relevance, and comparability in the realm of materiality. These guidelines would help to guide companies to effectively identify, evaluate, and prioritize material sustainability issues. Furthermore, stakeholders, including investors, customers, and regulatory bodies, can benefit from these results by gaining a deeper understanding of the transparency and diligence demonstrated by companies in communicating their materiality assessment processes. Companies that provide more detailed disclosures may receive favorable consideration from stakeholders with a keen interest in robust sustainability reporting. This, in turn, can contribute to fostering trust and confidence among stakeholders and potentially enhance a company's reputation in the realm of sustainability practices.

Functioning as both a compass and a guide, the book yields insights into the existing body of knowledge while also delineating pathways for future exploration to guide the trajectory of sustainability reporting studies. First, there emerges the need for research on corporate culture, in order to explore the role of non-financial disclosure to strengthen corporate commitment to sustainability and performance issues rather than just a set of burdens and requirements. Furthermore, there is still disagreement in the literature of a positive relationship between the quality of sustainability reporting and financial performance, even if the majority of existing research tend to suggest that high-quality disclosures somehow benefit the market and the reporting entities. Future research may further explore the role of assurance for sustainability reporting, to investigate under which conditions this activity improves the quality of sustainability reporting and its credibility for users.

An additional area for future development is materiality. Empirical studies may investigate materiality assessment disclosure comparing companies from diverse sectors, sizes, and geographical settings. Moreover, a longitudinal approach could significantly enrich insights by capturing trends and changes over time and would allow researchers to observe the evolution of reporting practices and identify potential patterns. Furthermore, understanding what factors influence the transparency of materiality assessments and exploring the potential consequences of high or low transparency would be valuable areas for future research to enhance our comprehension of the underlying dynamics at play. Factors such as corporate governance structures, industry characteristics, and regulatory environments could be examined to understand the drivers of transparency in materiality assessments. Additionally, assessing the impacts of transparency on various stakeholders and corporate outcomes would be valuable. Understanding how the level of transparency correlates with stakeholder trust, financial performance, and overall sustainability practices could provide meaningful insights. Comparative studies across different reporting frameworks, such as GRI and SASB, would contribute to evaluating the effectiveness of various standards in shaping materiality disclosures and could highlight the strengths and weaknesses of different frameworks. Qualitative approaches, such as in-depth interviews or case studies, could complement quantitative analysis. These methods could provide a deeper understanding of the underlying rationale behind companies' quality of disclosure on materiality assessment processes, offering richer insights into contextual factors influencing disclosure decisions. Lastly, investigating stakeholder perceptions of materiality disclosures could be a valuable research avenue. Surveys or interviews with stakeholders could help gauge their expectations, preferences, and the perceived usefulness of materiality information in decision-making processes.

REFERENCES

Adams, C. A., & Larrinaga-González, C. (2007). Engaging with organisations in pursuit of improved sustainability accounting and performance. *Accounting, Auditing & Accountability Journal, 20*(3), 333–355.

Barth, M. E., Cahan, S. F., Chen, L., & Venter, E. R. (2017). The economic consequences associated with integrated report quality: Capital market and real effects. *Accounting, Organizations and Society, 62*, 43–64.

Beske, F., Haustein, E., & Lorson, P. C. (2020). Materiality analysis in sustainability and integrated reports. *Sustainability Accounting, Management and Policy Journal, 11*(1), 162–186.

Cho, C. H., Laine, M., Roberts, R. W., & Rodrigue, M. (2015a). Organized hypocrisy, organizational façades, and sustainability reporting. *Accounting, Organizations and Society, 40*, 78–94.

Cho, C. H., Michelon, G., Patten, D. M., & Roberts, R. W. (2015b). CSR disclosure: The more things change...? *Accounting, Auditing & Accountability Journal, 28*(1), 14–35.

Dilling, P. F., & Harris, P. (2018). Reporting on long-term value creation by Canadian companies: A longitudinal assessment. *Journal of Cleaner Production, 191*, 350–360.

Eccles, R. G., Krzus, M. P., Rogers, J., & Serafeim, G. (2012). The need for sector-specific materiality and sustainability reporting standards. *Journal of Applied Corporate Finance, 24*(2), 65–71.

Farneti, F., & Guthrie, J. (2009, June). Sustainability reporting by Australian public sector organisations: Why they report. *Accounting Forum, 33*(2), 89–98.

Farooq, M. B., & De Villiers. (2020). How sustainability assurance engagement scopes are determined, and its impact on capture and credibility enhancement. *Accounting, Auditing & Accountability Journal, 33*(2), 417–445.

Farooq, M. B., Zaman, R., Sarraj, D., & Khalid, F. (2021). Examining the extent of and drivers for materiality assessment disclosures in sustainability reports. *Sustainability Accounting, Management and Policy Journal, 12*(5), 965–1002.

Fasan, M., & Mio, C. (2017). Fostering stakeholder engagement: The role of materiality disclosure in integrated reporting. *Business Strategy and the Environment, 26*(3), 288–305.

García-Sánchez, I. M., Hussain, N., Aibar-Guzmán, C., & Aibar-Guzmán, B. (2022). Assurance of corporate social responsibility reports: Does it reduce decoupling practices? *Business Ethics, the Environment & Responsibility, 31*(1), 118–138.

Kim, J., Cho, K., & Park, C. K. (2019). Does CSR assurance affect the relationship between CSR performance and financial performance? *Sustainability, 11*(20), 5682.

Machado, B. A. A., Dias, L. C. P., & Fonseca, A. (2021). Transparency of materiality analysis in GRI-based sustainability reports. *Corporate Social Responsibility and Environmental Management, 28*(2), 570–580.

Maroun, W. (2019). Does external assurance contribute to higher quality integrated reports? *Journal of Accounting and Public Policy, 38*(4), 106670.

Michelon, G., Pilonato, S., & Ricceri, F. (2015). CSR reporting practices and the quality of disclosure: An empirical analysis. *Critical Perspectives on Accounting, 33*, 59–78.

Milne, M., & Gray, R. H. (2007). Future prospects for sustainability reporting. In *Sustainability accounting and accountability* (pp. 184–208). Routledge Taylor & Francis Group.

Milne, M. J., Tregidga, H., & Walton, S. (2009). Words not actions! The ideological role of sustainable development reporting. *Accounting, Auditing & Accountability Journal, 22*(8), 1211–1257.

Puroila, J., & Mäkelä, H. (2019). Matter of opinion: Exploring the socio-political nature of materiality disclosures in sustainability reporting. *Accounting, Auditing & Accountability Journal, 32*(4), 1043–1072.

Ruiz-Lozano, M., De Vicente-Lama, M., Tirado-Valencia, P., & Cordobés-Madueño, M. (2022). The disclosure of the materiality process in sustainability reporting by Spanish state-owned enterprises. *Accounting, Auditing & Accountability Journal, 35*(2), 385–412.

Steenkamp, N. (2018). Top ten South African companies' disclosure of materiality determination process and material issues in integrated reports. *Journal of Intellectual Capital, 19*(2), 230–247.

Stolowy, H., & Paugam, L. (2018). The expansion of non-financial reporting: An exploratory study. *Accounting and Business Research, 48*(5), 525–548.

Stolowy, H., & Paugam, L. (2023). Sustainability reporting: Is convergence possible? *Accounting in Europe, 20*(2), 139–165.

Thorne, L., Mahoney, L. S., & Manetti, G. (2014). Motivations for issuing standalone CSR reports: A survey of Canadian firms. *Accounting, Auditing & Accountability Journal, 27*(4), 686–714.

REFERENCES

AAA (American Accounting Association). (1971). Report of the committee on non-financial measures of effectiveness. *The Accounting Review* (Suppl. to 46), 165–211.

AAA (American Accounting Association). (1975). Report of the committee on accounting for social performance. *The Accounting Review* (Suppl. to 50), 38–69.

AAA (American Accounting Association). (1978). *Report of the committee on the social consequences of accounting information.* AAA.

AASB (Australian Accounting Standards Board). (1990). *Qualitative characteristics of financial information.* Statement of Accounting Concepts No. 3. ASCPA.

Abhayawansa, S. (2022). Swimming against the tide: Back to single materiality for sustainability reporting. *Sustainability Accounting, Management and Policy Journal, 13*(6), 1361–1385.

Abhayawansa, S., & Adams, C. (2022). Towards a conceptual framework for non-financial reporting inclusive of pandemic and climate risk reporting. *Meditari Accountancy Research, 30*(3), 710–738.

Accountability. (2013). *Redefining materiality II: Why it matters, who's involved, and what it means for corporate leaders and boards.* https://lifegateedu.it/wp-content/uploads/2021/09/AA_Materiality_Report_Aug2013-FINAL_compressed.pdf

Accountability. (2018). *AA1000 accountability principles.* https://www.accountability.org/static/6b3863943105f2a5c4d5fc96affb750d/aa1000_accountability_principles_2018.pdf

C. Mio et al., *Sustainability Reporting*, Palgrave Studies in Impact Finance, https://doi.org/10.1007/978-3-031-58449-7

Adams, C. A. (2002). Internal organisational factors influencing corporate social and ethical reporting: Beyond current theorising. *Accounting, Auditing & Accountability Journal, 15*(2), 223–250.

Adams, C. A., & Larrinaga-González, C. (2007). Engaging with organisations in pursuit of improved sustainability accounting and performance. *Accounting, Auditing & Accountability Journal, 20*(3), 333–355.

Adams, C. A., Potter, B., Singh, P. J., & York, J. (2016). Exploring the implications of integrated reporting for social investment (disclosures). *The British Accounting Review, 48*(3), 283–296.

Adler, R., Mansi, M., & Pandey, R. (2018). Biodiversity and threatened species reporting by the top Fortune Global companies. *Accounting, Auditing & Accountability Journal, 31*(3), 787–825.

Adler, R., Mansi, M., Pandey, R., & Stringer, C. (2017). United Nations Decade on biodiversity: A study of the reporting practices of the Australian mining industry. *Accounting, Auditing & Accountability Journal, 30*(8), 1711–1745.

Agostini, M., & Costa, E. (2018). Financial and sustainability reporting: An empirical investigation of their relationship in the Italian context. *Sustainability and Social Responsibility: Regulation and Reporting*, 411–441.

Agostini, M., Costa, E., & Korca, B. (2022). Non-financial disclosure and corporate financial performance under directive 2014/95/EU: Evidence from Italian listed companies. *Accounting in Europe, 19*(1), 78–109.

AICPA (American Institute of Certified Public Accountants). (1973). *Objectives of financial statements*. AICPA.

AICPA (American Institute of Certified Public Accountants). (1976). *The measurement of corporate social performance*. AICPA.

Al Hawaj, A. Y., & Buallay, A. M. (2022). A worldwide sectorial analysis of sustainability reporting and its impact on firm performance. *Journal of Sustainable Finance & Investment, 12*(1), 62-86.

Aureli, S., Magnaghi, E., & Salvatori, F. (2019). The role of existing regulation and discretion in harmonising non-financial disclosure. *Accounting in Europe, 16*(3), 290–312.

Barbu, E. M., Ionescu-Feleagǎ, L., & Ferrat, Y. (2022). The evolution of environmental reporting in Europe: The role of financial and non-financial regulation. *The International Journal of Accounting, 57*(02), 2250008.

Barkemeyer, R., Preuss, L., & Lee, L. (2015, December). Corporate reporting on corruption: An international comparison. *Accounting Forum, 39*(4), 349–365.

Barnett, M. L. (2007). Stakeholder influence capacity and the variability of financial returns to corporate social responsibility. *Academy of Management Review, 32*(3), 794–816.

Barone, E., Ranamagar, N., & Solomon, J. F. (2013, September). A Habermasian model of stakeholder (non) engagement and corporate (ir) responsibility reporting. *Accounting Forum, 37*(3), 163–181.

Barth, M. E., Cahan, S. F., Chen, L., & Venter, E. R. (2017). The economic consequences associated with integrated report quality: Capital market and real effects. *Accounting, Organizations and Society, 62*, 43–64.

Baumüller, J., & Schaffhauser-Linzatti, M.-M. (2018). In search of materiality for nonfinancial information—Reporting requirements of the Directive 2014/95/EU. *NachhaltigkeitsManagementForum|Sustainability Management Forum, 26*(1–4), 101–111.

Baumüller, J., & Sopp, K. (2022). Double materiality and the shift from non-financial to European sustainability reporting: Review, outlook and implications. *Journal of Applied Accounting Research, 23*(1), 8–28.

Bebbington, J., Higgins, C., & Frame, B. (2009). Initiating sustainable development reporting: Evidence from New Zealand. *Accounting, Auditing & Accountability Journal, 22*(4), 588–625.

Bebbington, J., Kirk, E., & Larrinaga, C. (2012). The production of normativity: A comparison of reporting regimes in Spain and the UK. *Accounting, Organizations and Society, 37*(2), 78–94.

Bebbington, J., & Larrinaga, C. (2014). Accounting and sustainable development: An exploration. *Accounting, Organizations and Society, 39*(6), 395–413.

Bellucci, M., Acuti, D., Simoni, L., & Manetti, G. (2021). Restoring an eroded legitimacy: The adaptation of nonfinancial disclosure after a scandal and the risk of hypocrisy. *Accounting, Auditing & Accountability Journal, 34*(9), 195–217.

Bellucci, M., Simoni, L., Acuti, D., & Manetti, G. (2019). Stakeholder engagement and dialogic accounting: Empirical evidence in sustainability reporting. *Accounting, Auditing & Accountability Journal, 32*(5), 1467–1499.

Bernstein, L. A. (1967). The concept of materiality. *The Accounting Review, 42*(1), 86–95.

Berthelot, S., Coulmont, M., & Serret, V. (2012). Do investors value sustainability reports? A Canadian study. *Corporate Social Responsibility and Environmental Management, 19*(6), 355–363.

Beske, F., Haustein, E., & Lorson, P. C. (2020). Materiality analysis in sustainability and integrated reports. *Sustainability Accounting, Management and Policy Journal, 11*(1), 162–186.

Biondi, L., Dumay, J., & Monciardini, D. (2020). Using the international integrated reporting framework to comply with EU directive 2014/95/EU: Can we afford another reporting façade? *Meditari Accountancy Research, 28*(5), 889–914.

Blacconiere, W. G., & Patten, D. M. (1994). Environmental disclosures, regulatory costs, and changes in firm value. *Journal of Accounting and Economics, 18*(3), 357–377.

Boiral, O., Heras-Saizarbitoria, I., & Brotherton, M. C. (2020). Professionalizing the assurance of sustainability reports: The auditors' perspective. *Accounting, Auditing & Accountability Journal, 33*(2), 309–334.

Bold, F. (2017). Compliance and reporting under the EU non-financial reporting directive: Requirements and opportunities, Czech Republic, Brussels, Brno.

Bolt, R., & Tregidga, H. (2023). Methodological Insights "Materiality is …": Sensemaking and sensegiving through storytelling. *Accounting, Auditing & Accountability Journal, 36*(1), 403–427.

Bouten, L., & Everaert, P. (2015). Social and environmental reporting in Belgium: 'Pour vivre heureux, vivons cachés.' *Critical Perspectives on Accounting, 33*, 24–43.

Bouten, L., Everaert, P., Van Liedekerke, L., De Moor, L., & Christiaens, J. (2011). Corporate social responsibility reporting: A comprehensive picture? *Accounting Forum, 35*(3), 187–204.

Bowen, R. M., Davis, A. K., & Matsumoto, D. A. (2005). Emphasis on pro forma versus GAAP earnings in quarterly press releases: Determinants, SEC intervention, and market reactions. *The Accounting Review, 80*(4), 1011–1038.

Braam, G. J., De Weerd, L. U., Hauck, M., & Huijbregts, M. A. (2016). Determinants of corporate environmental reporting: The importance of environmental performance and assurance. *Journal of Cleaner Production, 129*, 724–734.

Brammer, S., & Pavelin, S. (2008). Factors influencing the quality of corporate environmental disclosure". *Business Strategy and the Environment, 17*(2), 120–136.

Brennan, N., & Gray, S. (2005). The impact of materiality: Accounting's best kept secret. *Asian Academy of Management Journal of Accounting and Finance, 1*, 1–31.

Brown, J. (2009). Democracy, sustainability and dialogic accounting technologies: Taking pluralism seriously. *Critical Perspectives on Accounting, 20*(3), 313–342.

Brown, J., & Dillard, J. (2013). Agonizing over engagement: SEA and the "death of environmentalism" debates. *Critical Perspectives on Accounting, 24*(1), 1–18.

Brown, J., & Dillard, J. (2014). Integrated reporting: On the need for broadening out and opening up. *Accounting, Auditing & Accountability Journal, 27*(7), 1120–1156.

Brown, J., & Dillard, J. (2015). Dialogic accountings for stakeholders: On opening up and closing down participatory governance. *Journal of Management Studies, 52*(7), 961–985.

Brunsson, N. (2007). *The consequences of decision-making.* Oxford University Press.

Buhr, N. (1998). Environmental performance, legislation and annual report disclosure: The case of acid rain and Falconbridge. *Accounting, Auditing & Accountability Journal, 11*(2), 163–190.

Buhr, N. (2002). A structuration view on the initiation of environmental reports. *Critical Perspectives on Accounting, 13*(1), 17–38.

Calabrese, A., Costa, R., & Rosati, F. (2015). A feedback-based model for CSR assessment and materiality analysis. *Accounting Forum, 39*(4), 312–327.

Camodeca, R., Almici, A., & Sagliaschi, U. (2018). Sustainability disclosure in integrated reporting: Does it matter to investors? A cheap talk approach. *Sustainability, 10*(12), 4393.

Canning, M., O'Dwyer, B., & Georgakopoulos, G. (2019). Processes of auditability in sustainability assurance—The case of materiality construction. *Accounting and Business Research, 49*(1), 1–27.

Carvajal, M., & Nadeem, M. (2023). Financially material sustainability reporting and firm performance in New Zealand. *Meditari Accountancy Research, 31*(4), 938–969.

Cerbone, D., & Maroun, W. (2020). Materiality in an integrated reporting setting: Insights using an institutional logics framework. *The British Accounting Review, 52*(3), 100876.

Chaidali, P., & Jones, M. J. (2017). It's a matter of trust: Exploring the perceptions of integrated reporting preparers. *Critical Perspectives on Accounting, 48*, 1–20.

Chiba, S., Talbot, D., & Boiral, O. (2018). Sustainability adrift: An evaluation of the credibility of sustainability information disclosed by public organizations. *Accounting Forum, 42*(4), 328–340.

Cho, C. H., Laine, M., Roberts, R. W., & Rodrigue, M. (2015a). Organized hypocrisy, organizational façades, and sustainability reporting. *Accounting, Organizations and Society, 40*, 78–94.

Cho, C. H., Michelon, G., & Patten, D. M. (2012). Impression management in sustainability reports: An empirical investigation of the use of graphs. *Accounting and the Public Interest, 12*(1), 16–37.

Cho, C. H., Michelon, G., Patten, D. M., & Roberts, R. W. (2015b). CSR disclosure: The more things change...? *Accounting, Auditing & Accountability Journal, 28*(1), 14–35.

Cho, C. H., & Patten, D. M. (2013). Green accounting: Reflections from a CSR and environmental disclosure perspective. *Critical Perspectives on Accounting, 24*(6), 443–447.

Chong, H. G. (2015). A review on the evolution of the definitions of materiality. *International Journal of Economics and Accounting, 6*(1), 15.

Christ, K. L., Rao, K. K., & Burritt, R. L. (2019). Accounting for modern slavery: An analysis of Australian listed company disclosures. *Accounting, Auditing & Accountability Journal, 32*(3), 836–865.

Christensen, H. B., Hail, L., & Leuz, C. (2021). Mandatory CSR and sustainability reporting: Economic analysis and literature review. *Review of Accounting Studies, 26*(3), 1176–1248.

Clark, C. E. (2021). How do standard setters define materiality and why does it matter? *Business Ethics, the Environment & Responsibility, 30*(3), 378–391.

Clarkson, P. M., Li, Y., Richardson, G. D., & Vasvari, F. P. (2008). Revisiting the relation between environmental performance and environmental disclosure: An empirical analysis. *Accounting, Organizations and Society, 33*(4–5), 303–327.

Clarkson, P. M., Overell, M. B., & Chapple, L. (2011). Environmental reporting and its relation to corporate environmental performance. *Abacus, 47*(1), 27–60.

Clarkson, P., Li, Y., Richardson, G., & Tsang, A. (2019). Causes and consequences of voluntary assurance of CSR reports: International evidence involving Dow Jones Sustainability Index Inclusion and Firm Valuation. *Accounting, Auditing & Accountability Journal, 32*(8), 2451–2474.

COM(2001)366. Commission of the European Communities. (2001, July 18). Green paper—Promoting a European framework for corporate social responsibility. Bruxelles.

Comyns, B., & Figge, F. (2015). Greenhouse gas reporting quality in the oil and gas industry: A longitudinal study using the typology of "search", "experience" and "credence" information. *Accounting, Auditing & Accountability Journal, 28*(3), 403–433.

Comyns, B., Figge, F., Hahn, T., & Barkemeyer, R. (2013). Sustainability reporting: The role of "Search", "Experience" and "Credence" information. *Accounting Forum, 37*(3), 231–243.

Consolandi, C., Eccles, R. G., & Gabbi, G. (2022). How material is a material issue? Stock returns and the financial relevance and financial intensity of ESG materiality. *Journal of Sustainable Finance & Investment, 12*(4), 1045–1068.

Cooper, S., & Michelon, G. (2022). Conceptions of materiality in sustainability reporting frameworks: Commonalities, differences and possibilities. In C. Adams (A c. Di), *Handbook of accounting and sustainability* (pp. 44–66). Edward Elgar Publishing.

Cormier, D., & Gordon, I. M. (2001). An examination of social and environmental reporting strategies. *Accounting, Auditing & Accountability Journal, 14*(5), 587–617.

Corporate Reporting Dialogue. (2016). *Statement of common principles of materiality of the corporate reporting dialogue.* http://sirse.info/wp-content/upl oads/2016/04/Statement-of-Common-Principles-of-Materiality.pdf

Costa, E., & Agostini, M. (2016). Mandatory disclosure about environmental and employee matters in the reports of Italian-listed corporate groups. *Social and Environmental Accountability Journal, 36*(1), 10–33.

Crane, A., & Ruebottom, T. (2011). Stakeholder theory and social identity: Rethinking stakeholder identification. *Journal of Business Ethics, 102*, 77–87.

CSRD. (2022). Directive 2022/2464/EU of the European Parliament and of the Council of 14 December 2022 amending Regulation (EU) No 537/2014, Directive 2004/109/EC, Directive 2006/43/EC and Directive 2013/34/EU, as regards corporate sustainability reporting.

Cuadrado-Ballesteros, B., Martínez-Ferrero, J., & García-Sánchez, I. M. (2017). Mitigating information asymmetry through sustainability assurance: The role of accountants and levels of assurance. *International Business Review, 26*(6), 1141–1156.

De Klerk, M., De Villiers, C., & Van Staden, C. (2015). The influence of corporate social responsibility disclosure on share prices: Evidence from the United Kingdom. *Pacific Accounting Review, 27*(2), 208–228.

De Villiers, C., & Alexander, D. (2014). The institutionalisation of corporate social responsibility reporting. *The British Accounting Review, 46*(2), 198–212.

De Villiers, C., La Torre, M., & Molinari, M. (2022). The Global Reporting Initiative's (GRI) past, present and future: Critical reflections and a research agenda on sustainability reporting (standard-setting). *Pacific Accounting Review, 34*(5), 728–747.

Deegan, C. (2002). Introduction: The legitimising effect of social and environmental disclosures–a theoretical foundation. *Accounting, auditing & accountability journal, 15*(3), 282–311.

Deegan, C. and Unerman, J. eds., (2011). Financial Accounting Theory. 2nd ed. Berkshire: McGraw-Hill.

Deegan, C. (2017). Twenty-five years of social and environmental accounting research within critical perspectives of accounting: Hits, misses and ways forward. *Critical Perspectives on Accounting, 43*, 65–87.

Deegan, C., & Rankin, M. (1997). The materiality of environmental information to users of annual reports. *Accounting, Auditing & Accountability Journal, 10*(4), 562–583.

Deegan, C., Rankin, M., & Tobin, J. (2002). An examination of the corporate social and environmental disclosures of BHP from 1983–1997: A test of legitimacy theory. *Accounting, Auditing & Accountability Journal, 15*(3), 312–343.

Delbard, O. (2008). CSR legislation in France and the European regulatory paradox: An analysis of EU CSR policy and sustainability reporting practice. *Corporate Governance, 8*, 397–405.

Dillard, J., & Roslender, R. (2011). Taking pluralism seriously: Embedded moralities in management accounting and control systems. *Critical Perspectives on Accounting, 22*(2), 135–147.

Dilling, P. F., & Harris, P. (2018). Reporting on long-term value creation by Canadian companies: A longitudinal assessment. *Journal of Cleaner Production, 191*, 350–360.

Denscombe, M. (1998). *The good research guide for small-scale social research project.* Open University Press.

DiMaggio, P. J., & Powell, W. W. (1983). The iron cage revisited: Institutional isomorphism and collective rationality in organizational fields. *American Sociological Review*, 147–160.

Dye, R. A. (1985). Disclosure of nonproprietary information. *Journal of Accounting Research*, 123–145.

Doshi, A. R., Dowell, G. W., & Toffel, M. W. (2013). How firms respond to mandatory information disclosure. *Strategic Management Journal, 34*(10), 1209–1231.

Dumay, J., La Torre, M., & Farneti, F. (2019). Developing trust through stewardship: Implications for intellectual capital, integrated reporting, and the EU directive 2014/95/EU. *Journal of Intellectual Capital, 20*(1), 11–39.

Dumitru, M., Dyduch, J., Guşe, R., & Krasodomska, J. (2017). Corporate reporting practices in Poland and Romania–an ex-ante study to the new non-financial reporting European directive. *Accounting in Europe, 14*(3), 279–304.

EC—European Commission. (2011). *Summary report on the responses received to public consultation on disclosure of non-financial information by companies.* http://ec.europa.eu/finance/consultations/2010/non-financial-report ing/index_en.htm

Eccles, R. G., Krzus, M. P., Rogers, J., & Serafeim, G. (2012). The need for sector-specific materiality and sustainability reporting standards. *Journal of Applied Corporate Finance, 24*(2), 65–71.

Edgley, C. (2014). A genealogy of accounting materiality. *Critical Perspectives on Accounting, 25*(3), 255–271.

Edgley, C., Jones, M. J., & Atkins, J. (2015). The adoption of the materiality concept in social and environmental reporting assurance: A field study approach. *The British Accounting Review, 47*(1), 1–18.

EFRAG & GRI. (2023). *EFRAG-GRI joint statement of interoperability.*

Erkens, M., Paugam, L., & Stolowy, H. (2015). Non-financial information: State of the art and research perspectives based on a bibliometric study. *Comptabilité-Contrôle-Audit, 21*(3), 15–92.

EU—European Union. (2014). *Directive as regards disclosure of non-financial and diversity information by certain large undertakings and groups, 2014/95/EU.*

European Commission. (2017). *Communication from the commission. Guidelines on non-financial reporting (methodology for reporting non-financial information) (2017/C 215/01)*. https://eur-lex.europa.eu/legal-content/EN/TXT/PDF/?uri=CELEX:52017XC0705(01)

European Commission. (2019). *Communication from the commission. Guidelines on non-financial reporting: Supplement on reporting climate-related information (2019/C 209/01)*. https://eur-lex.europa.eu/legal-content/EN/TXT/PDF/?uri=CELEX:52019XC0620(01)

Farneti, F., & Guthrie, J. (2009, June). Sustainability reporting by Australian public sector organisations: Why they report. *Accounting Forum, 33*(2), 89–98.

Farooq, M. B., & De Villiers, C. (2019). Understanding how managers institutionalise sustainability reporting: Evidence from Australia and New Zealand. *Accounting, Auditing & Accountability Journal, 32*(5), 1240–1269.

Farooq, M. B., & De Villiers. (2020). How sustainability assurance engagement scopes are determined, and its impact on capture and credibility enhancement. *Accounting, Auditing & Accountability Journal, 33*(2), 417–445.

Farooq, M. B., Zaman, R., Sarraj, D., & Khalid, F. (2021). Examining the extent of and drivers for materiality assessment disclosures in sustainability reports. *Sustainability Accounting, Management and Policy Journal, 12*(5), 965–1002.

Fasan, M., & Mio, C. (2017). Fostering stakeholder engagement: The role of materiality disclosure in integrated reporting. *Business Strategy and the Environment, 26*(3), 288–305.

Financial Accounting Standards Board (FASB). (1975a). *FASB Statement No. 5, accounting for contingencies*. FASB.

Financial Accounting Standards Board (FASB). (1975b). *Discussion memorandum: Criteria for determining materiality*. FASB.

Financial Accounting Standards Board (FASB). (1976). *FASB Interpretation No. 14, reasonable estimation of the amount of a loss*. FASB.

Financial Accounting Standards Board (FASB). (1991a). *Issue No. 89-13, accounting for the cost of asbestos removal*. FASB.

Financial Accounting Standards Board (FASB). (1991b). *Issue No. 90-8, capitalization of costs to treat environmental contaminations*. FASB.

Financial Accounting Standards Board (FASB). (1995). *Issue No. 93-5, accounting for environmental liabilities*. FASB.

Financial Accounting Standards Board (FASB). (2018). Statement of Financial Accounting Concepts No. 8. Conceptual Framework for Financial Reporting. *Chapter 3, Qualitative Characteristics of Useful Financial Information.*

Fiandrino, S., Tonelli, A., & Devalle, A. (2022). Sustainability materiality research: A systematic literature review of methods, theories and academic themes. *Qualitative Research in Accounting & Management, 19*(5), 665–695.

Freeman, R. E. (1984). *Strategic management: A stakeholder approach.* Cambridge University Press.

Freeman, R.E. (1984). *Strategic management: A stakeholder approach.* Prentice-Hall, Englewood Cliffs.

Freeman, R. E. (2010). *Strategic management: A stakeholder approach.* Cambridge University Press.

Frost, G., Jones, S., Loftus, J., & Van Der Laan, S. (2005). A survey of sustainability reporting practices of Australian reporting entities. *Australian Accounting Review, 15*(35), 89–96.

Fuhrmann, S., Ott, C., Looks, E., & Guenther, T. W. (2017). The contents of assurance statements for sustainability reports and information asymmetry. *Accounting and Business Research, 47*(4), 369–400.

Gamble, G. O., Hsu, K., Jackson, C., & Tollerson, C. D. (1996). Environmental disclosures in annual reports: An international perspective. *The International Journal of Accounting, 31*(3), 293–331.

García-Sánchez, I. M., Hussain, N., Aibar-Guzmán, C., & Aibar-Guzmán, B. (2022). Assurance of corporate social responsibility reports: Does it reduce decoupling practices? *Business Ethics, the Environment & Responsibility, 31*(1), 118–138.

Gerwanski, J., Kordsachia, O., & Velte, P. (2019). Determinants of materiality disclosure quality in integrated reporting: Empirical evidence from an international setting. *Business Strategy and the Environment, 28*(5), 750–770.

Giner, B., & Luque-Vílchez, M. (2022). A commentary on the "new" institutional actors in sustainability reporting standard-setting: A European perspective. *Sustainability Accounting, Management and Policy Journal, 13*(6), 1284–1309.

Global Reporting Initiative (GRI). (2021a). *GRI 1: Foundation 2021.* https://www.globalreporting.org/how-to-use-the-gri-standards/gri-standards-english-language/

Global Reporting Initiative (GRI). (2021b). *GRI 3 Material Topics 2021.* https://www.globalreporting.org/how-to-use-the-gri-standards/gri-standards-english-language/

Global Reporting Initiative (GRI) and Sustainability Accounting Standards Board (SASB). (2021). *A practical guide to sustainability reporting using GRI and SASB standards.* https://www.globalreporting.org/media/mlkjpn1i/gri-sasb-joint-publication-april-2021.pdf

Global Reporting Initiative (GRI). (2022). GRI Universal Standards 2021. Frequently Asked Questions (FAQs).

Gray, R. (2000). Current developments and trends in social and environmental auditing, reporting and attestation: A review and comment. *International Journal of Auditing, 4*(3), 247–268.

Gray, R. (2002). The social accounting project and accounting organizations and society privileging engagement, imaginings, new accountings and pragmatism over critique? *Accounting, Organizations and Society, 27*(7), 687–708.

Gray, R. (2010). Is accounting for sustainability actually accounting for sustainability... and how would we know? An exploration of narratives of organisations and the planet. *Accounting, Organizations and Society, 35*(1), 47–62.

Gray, R., Dey, C., Owen, D., Evans, R., & Zadek, S. (1997). Struggling with the praxis of social accounting: Stakeholders, accountability, audits and procedures. *Accounting, Auditing & Accountability Journal, 10*(3), 325–364.

Gray, R., Kouhy, R., & Lavers, S. (1995). Corporate social and environmental reporting: A review of the literature and a longitudinal study of UK disclosure. *Accounting, Auditing & Accountability Journal, 8*(2), 47–77.

Gray, I., & Manson, S. (2008). *The audit process*. Thomson Learning.

Grewal, J., Hauptmann, C., & Serafeim, G. (2021). Material sustainability information and stock price informativeness. *Journal of Business Ethics, 171*, 513–544.

GRI. (2013). G4 sustainability reporting guidelines: reporting principles and standard disclosures. *Global Reporting Initiative*, Amsterdam. Retrieved from: https://respect.international/g4-sustainability-reporting-guidelines-reporting-principles-and-standard-disclosures/.

GRI. (2023). *Universal Standards. Setting a new global benchmark for sustainability reporting. They are in effect for reporting from 1 January 2023*. https://www.globalreporting.org/standards/standards-development/universal-standards/

Guidry, R. P., & Patten, D. M. (2010). Market reactions to the first-time issuance of corporate sustainability reports: Evidence that quality matters. *Sustainability Accounting, Management and Policy Journal, 1*(1), 33–50.

Guthrie, J., & Farneti, F. (2008). GRI sustainability reporting by Australian public sector organizations. *Public Money and Management, 28*(6), 361–366.

Guthrie, J., Cuganesan, S., & Ward, L. (2008, March). Industry specific social and environmental reporting: The Australian Food and Beverage Industry. *Accounting forum, 32*(1), 1–15.

Hahn, R., & Kühnen, M. (2013). Determinants of sustainability reporting: A review of results, trends, theory, and opportunities in an expanding field of research. *Journal of Cleaner Production, 59*, 5–21.

Haller, A., Link, M., & Groß, T. (2017). The term 'non-financial information'—A semantic analysis of a key feature of current and future corporate reporting. *Accounting in Europe, 14*(3), 407–429.

Haque, S., Deegan, C., & Inglis, R. (2016). Demand for, and impediments to, the disclosure of information about climate change-related corporate governance practices. *Accounting and Business Research, 46*(6), 620–664.

Harper Ho, V. (2020). Non-financial reporting & corporate governance: Explaining American divergence & its implications for disclosure reform. *Accounting, Economics, and Law: A Convivium, 10*(2), 20180043.

Healy, P. M., & Palepu, K. G. (2001). Information asymmetry, corporate disclosure, and the capital markets: A review of the empirical disclosure literature. *Journal of Accounting and Economics, 31*(1–3), 405–440.

Holstrum, G. L., & Messier, W. F. (1982). A review and integration of empirical research on materiality. *Auditing: A Journal of Practice & Theory, 2*(1), 45–63.

Hopwood, A. G. (2009). Accounting and the environment. *Accounting, Organizations and Society, 34*(3–4), 433–439.

Hrasky, S. (2012). Carbon footprints and legitimation strategies: Symbolism or action? *Accounting, Auditing & Accountability Journal, 25*(1), 174–198.

Hsu, C.-W., Lee, W.-H., & Chao, W.-C. (2013). Materiality analysis model in sustainability reporting: A case study at Lite-On Technology Corporation. *Journal of Cleaner Production, 57*, 142–151.

International Accounting Standards Board (IASB). (2018). *Conceptual framework for financial reporting.* https://www.ifrs.org/content/dam/ifrs/publications/pdf-standards/english/2021/issued/part-a/conceptual-framework-for-financial-reporting.pdf

International Auditing and Assurance Standards Board (IAASB). (2022). *Handbook of international quality management, auditing, review, other assurance, and related services pronouncements.*

International Sustainability Standards Board (ISSB). (2023). *IFRS S1 general requirements for disclosure of sustainability-related financial information.* https://www.ifrs.org/issued-standards/ifrs-sustainability-standards-navigator/ifrs-s1-general-requirements/#about

International Integrated Reporting Council (IIRC). (2013). *Materiality background paper for IR.* https://www.integratedreporting.org/wp-content/uploads/2013/03/IR-Background-Paper-Materiality.pdf

IIRC (International Integrated Reporting Council). (2021, January). *International <IR> framework.*

International Integrated Reporting Council (IIRC). (2021). *International <IR> framework.* https://integratedreporting.ifrs.org/wp-content/uploads/2021/01/InternationalIntegratedReportingFramework.pdf

Jones, M. J. (2010, June). Accounting for the environment: Towards a theoretical perspective for environmental accounting and reporting. *Accounting Forum, 34*(2), 123–138.

Jones, P., Comfort, D., & Hillier, D. (2016). Materiality in corporate sustainability reporting within UK retailing: Materiality in sustainability reporting and UK retailers. *Journal of Public Affairs, 16*(1), 81–90.

Jørgensen, S., Mjøs, A., & Pedersen, L. J. T. (2022). Sustainability reporting and approaches to materiality: Tensions and potential resolutions. *Sustainability Accounting, Management and Policy Journal, 13*(2), 341–361.

Joseph, C., & Taplin, R. (2011, March). The measurement of sustainability disclosure: Abundance versus occurrence. In *Accounting Forum, 35*(1), 19–31.

JSE. (2022). *Leading the way for a better tomorrow JSE sustainability disclosure guidance.* JSE Group. https://www.jse.co.za/sites/default/files/media/doc uments/JSE%20Sustainability%20Disclosure%20Guidance%20June%202022. pdf

Khan, M., Serafeim, G., & Yoon, A. (2016). Corporate sustainability: First evidence on materiality. *The Accounting Review, 91*(6), 1697–1724.

Kim, J., Cho, K., & Park, C. K. (2019). Does CSR assurance affect the relationship between CSR performance and financial performance? *Sustainability, 11*(20), 5682.

KPMG. (2017). *The KPMG survey of corporate responsibility reporting 2017.* https://assets.kpmg/content/dam/kpmg/xx/pdf/2017/10/kpmg-survey-of-corporate-responsibility-reporting-2017.pdf

KPMG. (2022). *KPMG global survey of sustainability reporting 2022.* https://kpmg.com/it/it/home/insights/2022/12/kpmg-global-survey-of-sustainab ility-reporting-2022.html

Krippendorff, K. (2004). *Content analysis: An Introduction to its methodology.* Sage.

Kuh, T., Shepley, A., Bala, G., & Flowers, M. (2020). *Dynamic materiality: Measuring what matters.* https://ssrn.com/abstract=3521035

Kulkarni, S. P. (2000). Environmental ethics and information asymmetry among organizational stakeholders. *Journal of Business Ethics, 27*, 215–228.

La Torre, M., Sabelfeld, S., Blomkvist, M., Tarquinio, L., & Dumay, J. (2018). Harmonising non-financial reporting regulation in Europe: Practical forces and projections for future research. *Meditari Accountancy Research, 26*(4), 598–621.

La Torre, M., Sabelfeld, S., Blomkvist, M., & Dumay, J. (2020). Rebuilding trust: Sustainability and non-financial reporting and the European Union regulation. *Meditari Accountancy Research, 28*(5), 701–725.

Lai, A., Melloni, G., & Stacchezzini, R. (2017). What does materiality mean to integrated reporting preparers? An empirical exploration. *Meditari Accountancy Research, 25*(4), 533–552.

Leong, S., & Hazelton, J. (2019). Under what conditions is mandatory disclosure most likely to cause organisational change? *Accounting, Auditing & Accountability Journal, 32*(3), 811–835.

Lepoutre, J. M., & Valente, M. (2012). Fools breaking out: The role of symbolic and material immunity in explaining institutional nonconformity. *Academy of Management Journal, 55*(2), 285–313.

Lindblom, C. K. (1994). The implications of organizational legitimacy for corporate social performance and disclosure. In *Critical Perspectives on Accounting Conference, New York.*

Lin-Hi, N., & Müller, K. (2013). The CSR bottom line: Preventing corporate social irresponsibility. *Journal of Business Research, 66*(10), 1928–1936.

Littell, J. H., Corcoran, J., & Pillai, V. (2008). *Systematic reviews and meta-analysis.* Oxford University Press.

Lounsbury, M. (2008). Institutional rationality and practice variation: New directions in the institutional analysis of practice. *Accounting, Organizations and Society, 33*(4–5), 349–361.

Luque-Vilchez, M., & Larrinaga, C. (2016). Reporting models do not translate well: Failing to regulate CSR reporting in Spain. *Social and Environmental Accountability Journal, 36*(1), 56–75.

Machado, B. A. A., Dias, L. C. P., & Fonseca, A. (2021). Transparency of materiality analysis in GRI-based sustainability reports. *Corporate Social Responsibility and Environmental Management, 28*(2), 570–580.

Magness, V. (2006). Strategic posture, financial performance and environmental disclosure: An empirical test of legitimacy theory. *Accounting, Auditing & Accountability Journal, 19*(4), 540–563.

Mäkelä, H., & Näsi, S. (2010). Social responsibilities of MNCs in downsizing operations: A Finnish forest sector case analysed from the stakeholder, social contract and legitimacy theory point of view. *Accounting, Auditing & Accountability Journal, 23*(2), 149–174.

Malsch, B. (2013). Politicizing the expertise of the accounting industry in the realm of corporate social responsibility. *Accounting, Organizations and Society, 38*(2), 149–168.

Manetti, G. (2011). The quality of stakeholder engagement in sustainability reporting: Empirical evidence and critical points. *Corporate Social Responsibility and Environmental Management, 18*(2), 110–122.

Manetti, G., & Bellucci, M. (2016). The use of social media for engaging stakeholders in sustainability reporting. *Accounting, Auditing & Accountability Journal, 29*(6), 985–1011.

Maroun, W. (2019). Does external assurance contribute to higher quality integrated reports? *Journal of Accounting and Public Policy, 38*(4), 106670.

Martin-Sardesai, A., & Guthrie, J. (2019). Social report innovation: Evidence from a major Italian bank 2007–2012. *Meditari Accountancy Research, 28*(1), 72–88.

Martínez-Ferrero, J., & García-Sánchez, I. M. (2017). Sustainability assurance and cost of capital: Does assurance impact on credibility of corporate social responsibility information? *Business Ethics: A European Review, 26*(3), 223–239.

Massaro, M., Dumay, J., & Guthrie, J. (2016). On the shoulders of giants: Undertaking a structured literature review in accounting. *Accounting, Auditing & Accountability Journal, 29*(5), 767–801.

Mathews, M. R. (1997). Twenty-five years of social and environmental accounting research: Is there a silver jubilee to celebrate? *Accounting, Auditing and Accountability Journal, 10*(4), 481–531.

Melloni, G., Caglio, A., & Perego, P. (2017). Saying more with less? Disclosure conciseness, completeness and balance in Integrated Reports. *Journal of Accounting and Public Policy, 36*(3), 220–238.

Merkl-Davies, D. M., & Brennan, N. M. (2007). Discretionary disclosure strategies in corporate narratives: Incremental information or impression management? *Journal of Accounting Literature, 27*, 116–196.

Messier, W. F., Martinov-Bennie, N., & Eilifsen, A. (2005). A review and integration of empirical research on materiality: Two decades later. *Auditing: A Journal of Practice & Theory, 24*(2), 153–187.

Meyer, J. W., & Rowan, B. (1991). Institutionalized organizations: formal structure as myth and ceremony. *The new institutionalism in organizational analysis*, 41–62.

Michelon, G., Pilonato, S., & Ricceri, F. (2015). CSR reporting practices and the quality of disclosure: An empirical analysis. *Critical Perspectives on Accounting, 33*, 59–78.

Miles, S. (2017). Stakeholder theory classification: A theoretical and empirical evaluation of definitions. *Journal of Business Ethics, 142*, 437–459.

Miles, S., & Ringham, K. (2020). The boundary of sustainability reporting: Evidence from the FTSE100. *Accounting, Auditing & Accountability Journal, 33*(2), 357–390.

Millstone, S., & Watts, F. B. (1992). Effect of the green movement on business in the 1990's. *The Greening of American Business*, 1–31.

Milne, M., & Gray, R. H. (2007). Future prospects for sustainability reporting. In *Sustainability accounting and accountability* (pp. 184–208). Routledge Taylor & Francis Group.

Milne, M. J., & Gray, R. (2013). W (h) ither ecology? The triple bottom line, the global reporting initiative, and corporate sustainability reporting. *Journal of Business Ethics, 118*, 13–29.

Milne, M. J., Tregidga, H., & Walton, S. (2009). Words not actions! The ideological role of sustainable development reporting. *Accounting, Auditing & Accountability Journal, 22*(8), 1211–1257.

Mio, C., Agostini, M., & Panfilo, S. (2022). Bank risk appetite communication and risk taking: The key role of integrated reports. *Risk Analysis, 42*(3), 634–652.

Mio, C., Fasan, M., & Costantini, A. (2019). Materiality in integrated and sustainability reporting: A paradigm shift? *Business Strategy and the Environment, 29*(1), 306–320.

Mitchell, R. K., Agle, B. R., & Wood, D. J. (1997). Toward a theory of stakeholder identification and salience: Defining the principle of who and what really counts. *Academy of Management Review, 22*(4), 853–886.

Moggi, S. (2019). Social and environmental reports at universities: A Habermasian view on their evolution. *Accounting Forum, 43*(3), 283–326.

Moroney, R., & Trotman, K. T. (2016). Differences in auditors' materiality assessments when auditing financial statements and sustainability reports. *Contemporary Accounting Research, 33*(2), 551–575.

NAA (National Association of Accountants). (1974, February). Report to the committee on accounting for corporate social performance. *Management Accounting*, 39–41.

NFRD. (2014). Directive 2014/95/EU of the European Parliament and of the Council, 22 October 2014, amending Directive 2013/34/EU as regards disclosure of non-financial and diversity information by certain large undertakings and groups.

O'Connor, M., & Spangenberg, J. H. (2008). A methodology for CSR reporting: Assuring a representative diversity of indicators across stakeholders, scales, sites and performance issues. *Journal of Cleaner Production, 16*(13), 1399–1415.

O'Donovan, G. (2002). Environmental disclosures in the annual report. *Accounting, Auditing & Accountability Journal, 15*(3), 344–371.

Owen, D. L., Swift, T., & Hunt, K. (2001, September). Questioning the role of stakeholder engagement in social and ethical accounting, auditing and reporting. *Accounting Forum, 25*(3), 264–282.

Patten, D. M. (2002). The relation between environmental performance and environmental disclosure: A research note. *Accounting, Organizations and Society, 27*(8), 763–773.

Perera-Aldama, L. (2023). GRI and materiality: Discussions and challenges. *Sustainability Accounting, Management and Policy Journal, 14*(4), 884–903.

Pizzi, S., Caputo, F., & De Nuccio, E. (2024). Do sustainability reporting standards affect analysts' forecast accuracy? *Sustainability Accounting, Management and Policy Journal, ahead-of-print.* https://doi.org/10.1108/SAMPJ-04-2023-0227

Pizzi, S., Principale, S., & de Nuccio, E. (2022). Material sustainability information and reporting standards. Exploring the differences between GRI and SASB. *Meditari Accountancy Research, 31*(6), 1654–1674.

Plumlee, M., Brown, D., Hayes, R. M., & Marshall, R. S. (2015). Voluntary environmental disclosure quality and firm value: Further evidence. *Journal of Accounting and Public Policy, 34*(4), 336–361.

Puroila, J., & Mäkelä, H. (2019). Matter of opinion: Exploring the socio-political nature of materiality disclosures in sustainability reporting. *Accounting, Auditing & Accountability Journal, 32*(4), 1043–1072.

Qiu, Y., Shaukat, A., & Tharyan, R. (2016). Environmental and social disclosures: Link with corporate financial performance. *The British Accounting Review, 48*(1), 102–116.

Reimsbach, D., Schiemann, F., Hahn, R., & Schmiedchen, E. (2020). In the eyes of the beholder: Experimental evidence on the contested nature of materiality in sustainability reporting. *Organization & Environment, 33*(4), 624–651.

Rezaee, Z., & Tuo, L. (2019). Are the quantity and quality of sustainability disclosures associated with the innate and discretionary earnings quality? *Journal of Business Ethics, 155*(3), 763–786.

Rimmel, G., & Jonäll, K. (2013). Biodiversity reporting in Sweden: Corporate disclosure and preparers' views. *Accounting, Auditing & Accountability Journal, 26*(5), 746–778.

Roberts, R. W. (1992). Determinants of corporate social responsibility disclosure: An application of stakeholder theory. *Accounting, Organizations and Society, 17*(6), 595–612.

Roberts, R. W., & Dwyer, P. D. (1998). An analysis of materiality and reasonable assurance: Professional mystification and paternalism in auditing. *Journal of Business Ethics, 17*, 569–578.

Rodrigue, M., Magnan, M., & Cho, C. H. (2013). Is environmental governance substantive or symbolic? An empirical investigation. *Journal of Business Ethics, 114*, 107–129.

Ruiz-Lozano, M., De Vicente-Lama, M., Tirado-Valencia, P., & Cordobés-Madueño, M. (2022). The disclosure of the materiality process in sustainability reporting by Spanish state-owned enterprises. *Accounting, Auditing & Accountability Journal, 35*(2), 385–412.

Schiehll, E., & Kolahgar, S. (2021). Financial materiality in the informativeness of sustainability reporting. *Business Strategy and the Environment, 30*(2), 840–855.

Securities and Exchange Commission (SEC). (1999). *SEC staff accounting bulletin: No. 99—Materiality.* https://www.sec.gov/interps/account/sab99.htm

Securities and Exchange Commission (SEC). (2022). *SEC proposes rules to enhance and standardize climate-related disclosures for investors.* https://www.sec.gov/news/press-release/2022-46. Accessed 4 January 2024.

SEC (Securities and Exchange Commission). (2024). The Enhancement and Standardization of Climate-Related Disclosures for Investors, Release Nos. 33-11275; 34-99678; File No. S7-10-22, RIN 3235-AM87 (Washington D.C., March 6, 2024).

Sivarajah, U., Kamal, M. M., Irani, Z., & Weerakkody, V. (2017). Critical analysis of Big Data challenges and analytical methods. *Journal of Business Research, 70*, 263–286.

Steenkamp, N. (2018). Top ten South African companies' disclosure of materiality determination process and material issues in integrated reports. *Journal of Intellectual Capital, 19*(2), 230–247.

Stolowy, H., & Paugam, L. (2018). The expansion of non-financial reporting: An exploratory study. *Accounting and Business Research, 48*(5), 525–548.

Stolowy, H., & Paugam, L. (2023). Sustainability reporting: Is convergence possible? *Accounting in Europe, 20*(2), 139–165.

Suchman, M. C. (1995). Managing legitimacy: Strategic and institutional approaches. *Academy of Management Review, 20*(3), 571–610.

Sustainability Accounting Standards Board (SASB). (2017). *SASB conceptual framework.* https://www.sasb.org/wp-content/uploads/2019/05/SASB-Conceptual-Framework.pdf?source=post_page

Task Force on Climate-related Financial Disclosures (TCFD). (2017). *Recommendations of the task force on climate-related financial disclosures.* https://www.fsb-tcfd.org/recommendations/

Tate, W. L., Ellram, L. M., & Kirchoff, J. F. (2010). Corporate social responsibility reports: A thematic analysis related to supply chain management. *Journal of Supply Chain Management, 46*(1), 19–44.

Thorne, L., Mahoney, L. S., & Manetti, G. (2014). Motivations for issuing standalone CSR reports: A survey of Canadian firms. *Accounting, Auditing & Accountability Journal, 27*(4), 686–714.

Torelli, R., Balluchi, F., & Furlotti, K. (2019). The materiality assessment and stakeholder engagement: A content analysis of sustainability reports. *Corporate Social Responsibility and Environmental Management, 27*(2), 470–484.

Unerman, J., & Zappettini, F. (2014). Incorporating materiality considerations into analyses of absence from sustainability reporting. *Social and Environmental Accountability Journal, 34*(3), 172–186.

United Nations. (1992). *Environmental accounting, current issues, abstracts and bibliography.* United Nations.

Venturelli, A., Caputo, F., Cosma, S., Leopizzi, R., & Pizzi, S. (2017). Directive 2014/95/EU: Are Italian companies already compliant? *Sustainability, 9*(8), 1385.

Vitolla, F., & Raimo, N. (2018). Adoption of integrated reporting: Reasons and benefits—A case study analysis. *International Journal of Business and Management, 13*(12), 244–250.

Weerathunga, P. R., Xiaofang, C., Nurunnabi, M., Kulathunga, K. M. M. C. B., & Swarnapali, R. M. N. C. (2020). Do the IFRS promote corporate social responsibility reporting? Evidence from IFRS convergence in India. *Journal of International Accounting, Auditing and Taxation, 40*, 100336.

Whitehead, J. (2017). Prioritizing sustainability indicators: Using materiality analysis to guide sustainability assessment and strategy. *Business Strategy and the Environment, 26*(3), 399–412.

Windsor, D. (2017). Stakeholder responsibilities: Lessons for managers. In *Unfolding Stakeholder Thinking* (pp. 137–153). Routledge.

Wiseman, J. (1982). An evaluation of environmental disclosures made in corporate annual reports. *Accounting, Organizations, and Society, 7*(1), 53–63.

World Economic Forum. (2020). *Measuring stakeholder capitalism. Towards common metrics and consistent reporting of sustainable value creation.* https://www.weforum.org/publications/measuring-stakeholder-capitalism-towards-common-metrics-and-consistent-reporting-of-sustainable-value-creation/

INDEX

The manufacturer's authorised representative in the EU is Springer
Nature Customer Service Centre GmbH, Europaplatz 3, 69115 Heidelberg,
Germany. If you have any concerns regarding our products, please
contact ProductSafety@springernature.com

Printed and bound by CPI Group (UK) Ltd, Croydon, CR0 4YY
27/04/2026
02097562-0001